SACRED PLACES IN CHINA

CARL F. KUPFER

ZINC READ

Published by
This edition copyright © 2023 by Zinc Read
Design and setting By Zinc Read
Email- zincread@gmail.com
ISBN-978-93-5858-916-0

CONTENTS

PREFACE

The chief reason for producing this little volume is to give to the thoughtful reader of China and the Chinese a clearer conception of the readiness of the people to accept, with full credence, such whimsical and mythological stories as are here related, of their susceptibility of spiritual influences, and of the decay of intellectual vigor among the Buddhist and Taoist priests, as the inevitable result of monasticism. The intellectual vigor of the Chinese is found among the Confucianists, who hold the controlling power in the government, while Buddhism and Taoism seem past any hope of resurrection to real life. They have had their age of faith. But no one need to doubt the spiritual susceptibility, nor despair of the intellectual progress of all classes. Christianity fosters mental growth and science stimulates thought and is eminently fitted to drive out all fear and superstition. Christian education is not failing in accomplishing this. The response is abundantly gratifying. However, the struggle with Buddhism and Taoism is not yet ended, it has scarcely begun.

The reader finds himself here in the midst of the Asiatic world of nearly two thousand years ago, when Buddhist priests had entered actively upon their pilgrim life. To this day all foot-worn mountain paths lead to some monastery or sacred shrine.

The information recorded in this little volume is the fruit of hard labor. The writer traveled to distant mountains in the Mid-China hot summer months, visiting monasteries, and living with monks in the hope of gaining some knowledge of their inner life and hope of the future. Most of this information was obtained verbally, some through Chinese reading.

---

* * *

EARLY BUDDHISM IN HWANG MEI

It is probably not generally known, that in Hwang Mei Hsien, Hupeh, Buddhism found a most fruitful field in the early centuries of the Christian era. That long range of mountains north of Kiukiang is even now yet teeming with temples of more than ordinary fame. Thousands of pilgrims from many provinces visit these sacred shrines every autumn. The lamentable cries and moanings made by these pilgrims, when they prepare for starting on this pilgrimage, are not soon forgotten by those who spent the hot September nights in Central China cities. That their petitions may be answered, they take upon themselves the best vows, even promising not to look at a vain woman, while on the way to the temples.

The natural features of this granite mountain range are quite sufficiently interesting to explain why it was sought by devout devotees of early Buddhism. In some parts it is extremely rough and lies in such a rugged tumbled chaos, as scarcely to admit of classification. Although the general aspect is one of sheer desolation and barrenness, it must not be supposed that there is no fertility to be found. There are many rich valleys and pleasant little rivulets, fringed with verdure here and there, watering the rice fields far up the winding valleys; even the barest and most stony hillsides are seldom without vegetation.

The Fourth Patriarch's Monastery is situated at the upper end of one of these delightful valleys. Approaching it from the plain below, nothing is seen of it, until the bridge, spanning the *Pi Yü*, crystal stream, has been crossed, when the temple comes in view, standing in a large natural amphitheater, opening toward the south. By a remarkable coincidence, nature has here greatly aided the imagination and the credulous superstition of the Chinese mind. Two rugged spurs of mountains, supposed to be dragons, and as irregular as this fictitious animal can be, fighting for the *Chu* "Pearl," a little knoll in the open amphitheater, is certainly all that could be required, from the Chinese point of view, to make this a charming situation.

High up on a formidable mountain citadel, west of this monastery, stands a little temple, now in charge of a Taoist recluse. This temple is built over a natural cave in which, during the T'ang Dynasty, the Goddess of Mercy dwelt. When *Hwang Ch'ao*, the rebel, came to her abode in search of believers in her religion, she leaped across the deep ravine, which separated her from the monastery, and lit upon a solid granite rock, where her foot left an impress about three inches deep. Near by this footprint the place is seen where she rested. Rain and snow, frost and heat, drought and dampness of more than twelve hundred years have been unable to wash away the stain which the tears of the infuriated goddess left upon this granite rock.

The founder of this monastery, *Tao Hsin*, was born in *Honan* during the Northern *Chou* Dynasty in the reign of *Ta Hsiang*, A.D. 580. The Emperor, having heard of his precociousness and later of his virtues, sent him upon several occasions fine garments and invitations to come to the capital, but he always refused to accept the presents and declined to appear at court. When he was threatened with death if he persisted in his refusal, he calmly offered his head to the envoy. The Emperor, hearing that no Imperial consideration could induce him to improve his condition, left him in peace and built for him the *Pi Lu Ta*, a rafterless octangular temple. Approaching the monastery, this temple-tower is seen standing on the back of the dragon, which forms the left side of the amphitheater.

That this structure was built in the sixth century is quite credible; for it clearly shows age. The tooth of time has so deeply set its mark upon it, that its masonry resembles, in age, the rocks of the surrounding hills. No wood was used except for the two doors, one large front door and a small one at the east, which have, however, long since disappeared. Only one corner of the tower is shown on the photograph, the annex to the east is of modern work. In this temple the Patriarch lived and died in the latter part of the seventh century. Seven years after his death the doors burst open of their own accord, and it was seen that the body had not decayed. When the Emperor heard of this, he promoted him to a god among the deities of the Chinese Pantheon. In the *Ming* Dynasty, the fourth year of the Emperor *Wu Tsung*, A.D. 1520, it was seen that a flame of fire issued from his head and consumed his body. It is now replaced by a copper image, enshrined in the monastery within an inaccessible shrine. A dim light is kept burning before his gloomy, dismal altar.

In front of the main temple of this monastery stands a large cypress tree with the limbs grown downward, a strange freak of nature. The story told about it is that *Hung Jen* came to the Fourth Patriarch and desired to be instructed by him in the doctrine. The Patriarch replied, "You are too old now, you will have to wait for a new cycle." Thereupon the future successor of the Patriarchate pushed the small end of his staff into the ground and departed. The staff grew, and the result is the inverted cypress tree. Last year it died, root and branch. A telling illustration indeed of the once apparently flourishing and present decayed condition of this monastery and its inmates. The tree had not grown high. How could it with all its limbs turned down toward the earth! The marvel to us is that it grew as high as it did. What a struggle for life it must have had! And so the marvel to us is how such a class of men as the Buddhist mendicants are, lazy, filthy, ignorant parasites, have been able to maintain their position as leaders of a religion among a people so enlightened as the Chinese. A consciousness of extreme dearth of religious life among the people and no hope for improvement, or else no perception of spiritual need, is the only explanation why such leaders have been endured.

Fifteen *li* east of this monastery, in a valley even more beautiful and fertile than the one just described, richly watered by mountain streams, is the Temple of the Great Sun.

The founder of this temple was *Sung Ming I*, born in the *Kwangtung* province during the Eastern *Tsin* Dynasty, the fifth year of the Emperor *T'ai Yuan*, A.D. 381. He was, the incarnation of a dragon, who came to his virgin mother in a dream asking to be nursed. Her name was *Sing*. In the reign of *Sung An* he received the degree of Doctor of Philosophy. But when in office he found more pleasure in reading sacred books and charts than in his official duties. At an uprising among the people he resigned his position, hung his official seal in the judgment hall and gave himself to wandering about in the country, visiting temples. Traveling through *Kiangsi*, on a hot summer day, he was famishing of thirst. As he passed a man working in the field, he asked for some tea. The man replied, "I have no time, the sun will soon be down, and I must finish this work to-day." Said *Sung* to him, "If you will bring me some tea I will place my staff in the earth, and so long as it is there the sun will not go down." When he had refreshed himself with tea he walked off, leaving the staff standing in its place. The farmer, busy with his work, had not noticed the length of the day, until a neighbor came and spoke about the fixed sun; hearing of the reason he cut down the staff, and the spell was broken. Such is the story told and firmly believed by the people. In his journeyings he came to *Hwang Mei*. While there two young men associated themselves with him to be instructed by him in sacred learning.

At this time a long drought set in, and the people were in great need of water. He made his appearance before the people and announced that in seven days it would rain. So it did. And all the pools and streams were bountifully filled. After this he retired into seclusion, and lived in a mountain cave. When another drought came the people said to the *Hwang Mei* official, "Do nothing, but go up to the man living in the cave." The official went, and found him so intensely interesting that he forgot all about his own errand, and they began to write poetry and play chess. When he came out of the cave, behold, everything had changed. He knew no one, and no one knew him, nor could he even find his way back to *Hwang Mei.* Ages had come and gone; centuries had passed. So he returned to the cave and told Sung about it. While consulting together, sitting on a rock near the cave, they both with the rock ascended to heaven. On the stone, where last their feet rested, their footprints can yet be seen, also a chessboard carved into a stone in the cave. In the Temple of the Great Sun his image in life-size is the chief idol. Here, as in many other places in China, the people seemed to have more faith in this local deity than they have in the Buddhist gods. In connection with this temple, far up in the mountain in a quiet little nook, stands a plain little building, neatly kept, housing a large image of *Sung Ming I.* The old hermit in charge of this enviable little trust cultivates for his living a few little rice fields. But he has a pathetic story to tell, when we expressed our delight in finding such a cool retreat on a hot summer day. "Ah yes," said he, "what is joy to one brings sorrow to another; a few days ago, the last day in August, frost injured our rice crop."

Leaving the Great Sun Temple, we traveled across a range of most varied scenery. Having walked a distance of fifty *li*, and at an altitude of about three thousand feet, we suddenly came upon a large open valley, like a basin in the mountain top, surrounded by thickly wooded hills. Here nature had indeed been lavish in spending its beauty and glory, and here is located

The Old Patriarch's Monastery.

No wonder that the Patriarch was charmed with this situation at first sight. The story of his life was told us as follows: He was a native from Mid-India, born during the *Chou* Dynasty, in the reign of the Emperor *Wei Lei*, 425 B.C. From birth his left hand was clenched into a fist, until his mother for the first time took him to the idol, when his hand opened, so that he could fold his hands respectfully in worship, whereupon he became deeply interested in idols. Being an unusually precocious boy, he read widely until he came across the Buddhist Canon, and forthwith embraced that religion. After long and patient contemplation he came to China, and on his way north he met the sainted priest, *Ta Mo*, at the *Lü* Mountains near Kiukiang. The name of the place is called *Ho Shang Fên*, that is, the place where the priests separated. *Ta Mo* had come from Southern India, and was the last of the Western and the first of the Eastern Patriarchs of Buddhism. *Ta Mo* asked him how old he was. He replied, "My age is eight hundred years." "So," said *Ta Mo*, "I have been looking for a man of great age, you may be the Old Patriarch," and gave him the robe and bowl of the Patriarchate. After this conference at the foot of the *Lü* Mountains, the Old Patriarch crossed the swollen Yangtze, and chose the *Hwang Mei* district as his field of labor. Having set his heart upon the location for a temple, but finding no wood to build with, he went to *Lung Ping*, on the *Yangtze* River, where a merchant had a large raft of wood. Said he to the merchant, "May I sink this raft?" The merchant replied, "If you can sink that raft I will give it to you." The raft disappeared, and came up again in that high mountain valley. In like miraculous way his daily needs were supplied. When he had no rice, out of a little hole in a rock rice came.

When the needed oil and salt, from another rock oil and salt flowed. Alas, to-day this once renowned monastery shows no such signs of divine favor. A sickly, diseased looking young man, and a silly boy were the only priests in charge. Their merit is certainly not in works of purity and wisdom duly combined, for none of these virtues are found with them. What *Ta Mo* (the founder of this temple) taught has apparently perpetuated itself, and may account somewhat for the ignorance and apathy of Buddhist priests, viz., that religion is not to be learned from books, but that man should seek and find the Buddha in his own heart. Every department of this great labyrinthian temple shows utter neglect and rapid decay. Even the main altar with Shakyamuni, the god of all gods, Kaskiapa, the god of the future, and Ananda, the god of the past, are in a most dilapidated and pitiable condition.

It may be supposed that with some of these priests there is a desire for spiritual life, but undoubtedly their spiritual conception is of a very inferior order.

From the moment we entered this beautiful place in the mountain top until we took our departure the sacrilegious thought did not leave us, that it would make a most charming summer resort for Westerners. Traveling sixty *li* farther eastward over rugged hills and through valleys, along steep cliffs and jagged rocks, we came to

The Fifth Patriarch's Monastery.

The approach to this temple is a broad stone-paved road, rising immediately up from the level plain toward the massive cliff, like a huge altar of some natural temple, encircled by peaks of various shapes and heights. The founder of this monastery is claimed to have been born into this world the second time in the Sui Dynasty, and the second year of *Jen Chou*, A.D. 603. In his first stage of existence his name was *Hung Jen*, and he lived in *Hwang Mei*, where he was a planter of coniferous trees. He came to *Tao Hsin*, the Fourth Patriarch of Buddhism, for instruction. "May I yet learn the doctrine?" he asked. "Die, and come again, and I will teach you," replied *Tao Hsin*. *Hung Jen* asked, "Where shall I go to die?" "Go until you come to a pond where a maiden named *Chou*is washing clothes, and there stop," said *Tao Hsin.* He went, and when he came to the place and saw the woman, he said, "I am an old man, may I stop over night?" The maid replied, "I have a father and brother, ask them." While he yet spake a peach came up out of the water, and the maid ate it and she became his mother. Despised and rejected by her people she was driven away from her home, and with her son reduced to beggary. In dire distress she threw the child into a pond, but he would not sink. Rescued from this watery grave, she nursed him again; but the first seven years of his life he never spoke a word, and was mocked and ridiculed by many; saying, "He has neither father nor name." When his mother brought him the first time to the temple, the bells rang of their own accord. Hearing this he said, 'Now I can learn the doctrine." He went again to *Tao Hsin* to be instructed. "What is your name?" asked the Patriarch. "I have no name," was the reply. "What! no name?" "No," said he, "I am a Fu, an idol. I am emptiness." Having gained the favor of the Fourth Patriarch, he became his successor.

To the credit of *Hwang Mei* Buddhism, it may be said that this monastery is in a good state of repair, and favored with an intelligent Abbot. It would indeed have been a pleasure to have spent some time with him. He received us dressed in his priestly robes, and spoke very intelligently. But his embarrassment arose when we asked to have a look at the undecayed body of *Hung Jen.* The most liberal offer which we could make would not induce him to open the sacred shrine where this deified "Emptiness" rested. As with his predecessor, so here a dim light is kept burning before his altar.

When the philosophical and abstract teaching on speculative metaphysical subjects, such as the Patriarchs of Buddhism taught their disciples, are taken into account, then we need not be surprised at the indifference of the priests in their habits. Indeed, superficially viewed, the more slovenly they live the more orthodox they might seem in their own eyes. This may be inferred from examples of their doctrine. For instance, when *Hung Jen* desired to know who should be his successor, Giles tells us that he asked each of his monks to compose a gâthâ. One of his favorites wrote on a wall:

"Man's body is like the Bodhi trees;
His mind is like a mirror;
And should be constantly cleaned
Lest dust should stick to it."

Lu Hui-nêng saw this and came by night and wrote alongside:

"There is no such thing as the Bodhi trees;
There is no such thing as the mirror;

There is nothing, which has a real existence;
How then can dust be attracted?"

This secured for him the robe and bowl of the Patriarchate; for he became the last Patriarch of Buddhism in China, as the doctrine by him was supposed to be well established. From the above narrative the reader can conclusively notice that the Chinese folklore does not only teem with childish legends, but contains proofs of their implicit faith in miracles; belief in fanciful miracles without the acknowledgment of divine intervention. An unintelligent faith, which has never led them to search for the cause of the accepted wonders.

* * *

NGN REN TSZ—THE MONASTERY OF BENEVOLENCE

The premises of this temple are in close proximity to the campus of the Woman's Foreign Missionary Society at Kiukiang, only a narrow lane intervening. The temple facing south and the girls' high school north brings the two institutions face to face. Like many other renowned places in China this temple developed from a very humble beginning, a little shrine, until the virtue proceeding from it reached distant places, even to far-away Tibet. During the golden age of Buddhism, when the faith of that religion had a deep hold upon all classes, having even many officials among its devotees, an Abbot had a dream. In this dream Buddha appeared to him and informed him that on a certain day an incarnate being would be coming down the Yangtse River from Tibet. The Abbot at once notified the officials of the city of his wonderful dream, and elaborate preparations were made for the reception of this divinity. Upon the appointed time the officials all appeared in their official attire and together went to the north gate of the city, where they waited for the coming of the promised deity. Soon they saw, to their amazement, a stone boat floating down the river, and in it stood a majestic being. At the gate the boat drew ashore, and the man of the West, after a dangerous voyage and much fatigued, stepped off, and asked for Ngn Ren Tsz. With due ceremony and in great pomp he was escorted to the sought-for sacred shrine. The news of this strange visitor to this humble place spread throughout the provinces. Pilgrims came from all parts seeking the blessing which this divinity had brought to this temple. To assure the perpetuation of his favors, a life-sized metal image was made and enshrined as the patron of benevolence, and the temple flourished beyond all expectation.

The granite stone boat which brought this deity from the far West is lying in the temple grounds, and is proudly shown to this day by the credulous priests as the instrument that brought the once enjoyed fame to their temple. It is of one solid granite block about fifteen feet long and three feet wide, hollowed out like a trough. All admit that of itself it could not float, but the buoyant divinity bore it up, and it bore the divinity down through the gorges to his desired destination.

During the *Taiping* Rebellion this temple shared the same fate to which all pagan temples in Mid-China were doomed; all were destroyed. After the rebellion was over, and the Abbot returned to his sacred home, he found nothing but desolation and ruins. Out of the broken débris he constructed a little shed for his home and a shrine; but the former fame and virtue did not seem to return. No pilgrim worshipers sought his aid. Every effort to re-establish the past glory seemed in vain. Direst poverty stared him in the face. One day, as he was walking across the once beautiful grounds, he struck his foot against an object. With the usual epithet that always falls from the lips of a Chinese when something untoward happens, he was about to pass on, when, unlike his countrymen, he concluded to return and remove this obstacle, that no one else should henceforth be so unfortunate. When he began to dig with his mattock to remove the contemptible object, to his unspeakable surprise and joy he found it to be the image of the once famous iron Buddha, who had been the means of prosperity and fame in former times. The mutilated god was in a sad plight. One arm was entirely missing, and other serious defects were discovered. He handled him with great care. The missing hand was soon replaced, and all blemishes repaired. Newly gilded over and an altar extemporized, he was soon reinstated. When this was known the former reputation of the temple began to return. An Imperial grant was made for the erection of substantial buildings, and a new era for the once famous monastery seemed assured.

Twenty years ago it was doubtless one of the best kept pagan institutions in this part of China. It contained all of the departments necessary to constitute a fully equipped monastery. In a lofty building at the main entrance stand the four imposing images known as the Guardians. Then follows an artificially constructed pool over which is built the Bridge of Fate, leading to the Chief Hall, where the Buddha of the Past, Present, and Future, and the Eighteen Lohan are enshrined. On the north side of the center altar the Goddess of Mercy is represented as standing at the seashore upon the back of a great sea monster. Here, as in other parts of China, she is the most highly revered deity in the Buddhist pantheon. She is to the Buddhist what the Holy Mother Mary is to the Catholic. In the northeast corner of this building hangs a large bronze bell, whose melodious sound is not supposed to die out. Every stroke upon this bell with a horizontally suspended piece of wood drawn by a priest is believed to send a flash of light into the darkness of Buddhist hades, and illuminates the path for a struggling soul to escape from the torturing demons. The vibrations of this bell seem to continue long enough to allow the faithful priest to take a little nap between the strokes. For many years a blind priest sat there day and night performing this meritorious service.

In this spacious hall many priests meet to hold annual gatherings during the month of September. These meetings might be denominated

Liturgical conventions, or high mass. They are sometimes continued two or three weeks, and at the close young priests, having taken their vows, are consecrated to the priesthood by the Abbot, who, with a burning incense stick, brands them on top of their shaven heads.

In the extreme rear of these grounds is a modest little structure within which stands the once famous iron Buddha in a glass case, testifying of the fame of his past days. To the right and left of this building the monks have their living compartments.

To the east, close by the temple grounds, stands a pagoda for which unusual merit is claimed. Many years ago when the literati of Kiukiang had failed for successive years in their effort to pass in the competitive examination for degrees, one of their number, a *savant* above the average, advised that a pagoda be erected to regulate the *Fung shui*, that fortune might again return to the wise men of Kiukiang. All who had faith in this prophecy contributed freely, and enough energy was set in motion to so stimulate the sluggish literati that the very next examination, after the completion of the pagoda, brought brilliant success to a number of competitors. The merit of the pagoda was established beyond doubt, and every year thereafter students of *Teh Hwa-hsien* Township have been successful. At the last examination, given under the old régime, two of the students of William Nast College were among those who gained the enviable distinction. The competitive examination has been abolished, but the pagoda will doubtless stand as a monument of past superstition for ages to come.

The last effort to fan the dying embers of this decaying monastery by a sister religion was the erection of a Confucian Ancestorial Hall at the southwest corner, between the Home of the Woman's Foreign Missionary Society and the temple. The purpose of this effort is easily seen, even though Confucianists and Buddhists are not close friends. They are allied state religions, and will doubtless assist each other when the death knell of either is heard ringing. He who hath ears to hear can hear them ringing even now.

* * *

KIU HUA SHAN—OR THE NINE-LOTUS-FLOWER MOUNTAIN

The traveler on a Yangtze River steamer, passing from Wuhu to *Nganking*, will see about midway between these two cities, and twenty-five miles from the southern bank, a long range of mountains with unusually sharp, rugged peaks, the highest points of which are probably not over five thousand feet above the Yangtze Valley. The original name of this range was the "Nine Sons," but in the T'ang Dynasty the celebrated poet, *Li Pah*, made the observation that nine peaks were shaped like the lotus flower, hence it was called the Nine-Lotus-Flower Mountain. In the whole range there are ninety-nine peaks, the most conspicuous of which are:

Tien-T'ai, the Altar of Heaven.
Fung Yung, the Hibiscus Peak.
Chung-Fung, the Middle Peak.
Hui-Sien, the Peak of the Genii.
Wu-lao-fung, the Five Old Peaks.
Tien-chii, the Heavenly Pillar.
Cheu-ran-fung, the Wizard Peak.
Shuang-fung, the Twin Peaks.
Tu-hun, the Most Beautiful.

These peaks rise up keen and high, clothed in beautiful mantles of eternal verdure, holding their heads up in the pure atmosphere, far above the little petty trials and actions of men down in the valley. There may be mountains far more imposing, but for variety of scenery and richness of vegetation none can excel them. What a sight it must have been when they were formed, and who can weary praising their beauty and their greatness? We are not surprised that the heathen in his blindness seeks the home of his gods among these majestic granite peaks.

A mere glance at the richness of the flora of these mountains and valleys is of great interest to the lover of nature. We took note of only a few specimens, but one fact must be kept in mind, that much of the information obtained is tinged with mythological superstition.

Huang-li-ta, the yellow rice, which is said to grow high up the valleys. It was originally brought from Siam by Ti Tsang-Wang, being different from all other rice. It is very productive, of a reddish yellow color, fragrant, and soft.

Chu-shih, seed-producing bamboo. It happened twice during the *Ming* Dynasty that the bamboo on these hills produced seeds. In spring and summer they grow purple flowers, and in the autumn seed, like wheat.

Shih chih, the plant that contains the elixir of life. It grows high up on the steep cliffs, and to obtain it ropes must be let down from the top, upon which the adventurous climb up. When this plant is eaten, the body becomes light, as if it could fly, and long life is assured. The cliff upon which it grows is purple on the side facing the sun, and dark on the opposite side.

Chu tien, the bamboo mat, is soft and white like mushrooms. When cooked the juice from it is red like blood; the pulp is good for food, and of pleasant taste.

Chin ti tsa, the golden ground tea. The growth is hollow like the bamboo. The plant is supposed to have been brought from Siam by Ti Tsang Wang.

Ming Yuan Tsa, the *Ming* fountain tea, so called because it only grows at the *Ming* fountain, under the shadow of a hill. It does not put forth leaves until the end of spring and the beginning of summer. Its limbs are long, but do not spread out, and the color of its leaves changes suddenly from purple to green. Only the water of the *Ming* fountain can make good tea of these leaves.

Wu t'sai sung, the pinus massoniana, has a hair-pin shaped leaf. The seed of this tree is like small chestnuts, with three corners; the kernel is fragrant. The origin is uncertain, some claiming that it also was brought from Siam by *Ti Tsan Wang*.

Chien t'sai sung, the thousand autumn cedar, does not grow more than two or three inches high, and has leaves like the cypress. It is at home on the high, dry cliffs, and when apparently dead, if watered, it will revive and become green again, like the *huan hun t'sao*, the soul returning grass.

Sien jen chu, the sacred man's candle, is like the *wu t'ung* in color. When full grown it can be spanned with one hand, and is not over one foot high. It is so rich in sap that when dug up it will not dry within one year. The burning of one limb will give light for several hours. Dilettantes in fairy grottoes are supposed to use it for their lights when performing their pious frauds in literature.

Loh han poh, the Lohan cypress, thujadolabrata. It grows only one or two feet high, with leaves resembling the arbor vitae. The color is kingfisher green.

Fei hsien kai, the sacred flying cover. The trunk of this tree grows high and limbless. At the top only it has limbs growing down, like the ribs of an umbrella. It compares in beauty with the wistaria chinensis, the leaves being bright and variegated.

Lo han t'iao, the Lohan rope, grows high, and resembles the Chinese juniper. The leaves are purple red, mingled with other colors, having a white line in the center, and are over one foot long. When dry they roll up like a rope, hence the name *t'iao*.

P'u sa hsien, the idol thread, with an abundance of fine limbs and leaves drooping down almost to the ground, like threads. The leaves are fragrant and of a kingfisher green color.

Yü Nü Chang, the beautiful woman's curtain, is an evergreen, and grows ten to twenty feet high. The leaves are long, bright, and glossy, growing in circles, like a screen. It is found in the shade of the rocks and cliffs, often standing in rows.

Hsien chang fan, the sacred palm fan, grows large leaves out of the trunk, like the fans carried in processions. Two sides of the leaves are similar and regular, being of deep green color and fragrant. It is often found at the mouth of the caves.

Chin ch'ien shu, the golden cash tree, grows straight, twenty to thirty feet high. The trunk is dark, and the limbs and leaves are green. During summer and autumn fine thread-like limbs grow out of the larger limbs. Around these drooping threads are little circles like Chinese cash, some having from three to four, others as many as twelve such cash.

Nan t'ien Chu, the Southern heavenly bamboo, nandina domestica. Like all bamboo, it is hollow, and the nodes are far apart. In spring its color is green, in autumn and winter it is red and purple. It flourishes by the side of streams.

Kwan yin chu, the goddess of mercy bamboo, is of small growth, not quite three feet high. Its leaves are fine, and their color bluish green. It is an evergreen.

Shih chu, the stone bamboo, also called the dragon beard bamboo. It grows a fine flower, and its color is red and purple.

Chin pu yao, the shaking golden step. It grows in abundance and flowers profusely. Its leaves and flowers are always shaking, like the aspen.

Hsien kwei, the sacred cassia, has leaves fine and soft, of reddish and greenish colors. The color and shape of the flowers are like red sacks, and the seed like red pearl. When damp with dew both flowers and leaves emit a most pleasant fragrance.

Yü yin lo, the precious brooches. The flower is round and droops, like pure pearls. It grows along streams in clusters, like the garden thyme.

Po lan hua grows over ten feet high, with long fine leaves; the calyx is like the sunflower, and its fragrance is perceptible several *li* off. It is said that when *Ti Tsan Wang* went to the Southern altar one of these flowers fell into his patriarchal bowl. At other times this flower had never been known to fall.

So lo hua, a species of sedge. The growth is of different sizes, both large and small, quite different from all other plants. The leaves are always in clusters of seven or nine each. The seed in shaped like the face of a person, having eyes and eyebrows; and the flowers grow close together like the peony tree; their fragrance resembles that of the lotus flower.

La hua, the wax flower, is small, and always has fine, thick, soft petals, like wax. Its color is red and yellow.

Mo lien hua, a species of jasmine. The Chinese claim that it produces seed before it puts forth leaves. It blooms every month except November and December. In color and fragrance it resembles the lotus, shedding perfume on the passing air.

Shui hsien hua is like the jonquil or narcissus. It grows in dark places, and even opens its flowers while partly covered with snow.

Sung chu mei is a strange freak in nature. It grows like the cypress, but is hollow and has nodes like the bamboo and roots like the garlic. It blooms like the plumtree, but is poisonous.

Lung hu t'sao, the dragon beard grass, grows three feet high and has a fine stiff stem without nodes. It is found in abundance on the dizzy heights of the mountain peaks, and is used for making fine mats.

Kin chieh ch'ang pu, the nine-pointed calamus, is found in well-watered rocky places. The stem has nine nodes to one inch, and it is claimed that in the *T'ang* Dynasty some were seen with from twelve to twenty-four nodes to an inch. It is regarded as a sacred plant, and many Chinese poems have been written upon it. The leaves of this plant, with those of artemesia, are hung on the door lintels in the shape of a sword on the fifth day of the fifth moon; the Dragon Festival, in remembrance of the famous rebel, *Huang Ch'ao*, whose soldiers had orders to spare every family who exhibited a bunch of artemesia and calamus at the door. When it is eaten by aged people their gray hair will become black again.

"Floral apostles, that with dewy splendor
Blush without sin, and weep without a crime;
O! may I deeply learn, and ne'er surrender
Your love divine."

Fung wei t'sao, the male phoenix tail grass, is a fern with twin leaves diverging into two branches like the tail of the phoenix. It is claimed to possess cooling properties in medicine.

Chin hsing t'sao, the golden star plant, has golden specks on its leaves, and black fibres like hair within the stem.

Peh ho, the lily flower, also called the devil's garlic. The bulb can be eaten, and syrup is made from the juice.

Huang chin. The meaning of this name is yellow energy or spirit. It grows in many other places, but these mountains produce the finest quality. To eat the best quality prevents old age and disease, and assures long life.

"Hail, blessed flowers;
Springing in valleys green and low,
And on the mountain high;
And in the wilderness,
Where no man passes by."

It is not, however, the beautiful natural scenery, nor is it the richness of the flora of this mountain group that has made it famous among the mountains of China, but a great personage, whose influence was powerfully felt in shaping the Buddhist religion in China during the early centuries. Buddhist history claims this person to have been a prince of Siam, who, becoming weary of the pomp and vanity of court life, cultivated a love for the Buddhist religion. He was the son of the king. His surname was *Gold*, and he was called *Chiao Chio*. As a god he is called *Ti Tsan Wang*. In the T'ang Dynasty, during the reign of *Chih Tê*, A.D. 754, he forsook his luxurious home to live the life of a poor mendicant. Passing through this unknown country he was attracted by the grandeur of the mountains, and resolved to make them his home. He begged the magistrate of *Tsingyanghsien* to give him a plot of land in area as large as his coat would cover. The modest request was cheerfully granted, and he built a little stone hut. He lived upon rice and white clay, and drank the fresh, clear mountain water as an evidence of the purity of his desires.

His first adventure was to encounter a poisonous serpent. But the prince remained quietly seated and undisturbed in meditation, offering no resistance. Soon a beautiful woman came and made obeisance to him, apologizing for the baseness of the serpent's action, saying, "The child did not know what he was doing." As a compensation for the inconvenience the serpent had given him, she caused fresh water to bubble out of a rock near by, and to this day sparkling water flows from this rock. It is called the Dragon Daughter Spring, for the beautiful woman was the dragon's daughter, and the impudent serpent was her younger brother.

At *Tsingyanghsien*, a city near by, there lived a man named *Chü Kochieh*, who, passing over the hills one day, saw the prince sitting in a little stone hut with a tripod by his side, as a range for cooking his food. Seeing this he was astonished, and bought for him a piece of land, upon which he and *Shen Yü*, a pupil of the prince, built a temple for him. In later years this temple was named "*Hua Ch'en Tsz*" by Imperial permission.

Many of his Siamese countrymen, having heard of their prince's fame across the sea, followed him, and thus his disciples increased daily from abroad and at home. They lived with him a like frugal life, eating rice, millet, and white clay. When he had reached the age of ninety-nine years, and felt that he must soon depart, he called his disciples together. No sooner had they gathered around their beloved master than they heard prolonged, inarticulate sound reverberating among the mountains, and saw shafts of light scintillating along the horizon, until it seemed as if the hills cleft apart, when suddenly he sank into an open gap. They buried him where once stood the little stone hut, and every three years the sarcophagus was opened and his body was found as when living, his joints trembling and giving forth a rattling sound, like golden chains. Thus it was considered beyond all doubt that he had become a god, for flames of light issued from the place, and it was called "the mount of spiritual light." A pagoda is erected over this spot, and around the pagoda a temple is built. Here is the center of interest, for in this pagoda rests the undecayed body of *Ti Tsang Wang*. But to speak of the temples we must begin at the foot of the mountains.

Approaching from the *Tatung* plain, the first station, *Er Shen Tien*, Temple of the Holy One, is seen at a long distance. A night's lodging is here desirable, if the pilgrim's season—September, October, and November—is on. Bands of pilgrims, numbering from fifteen to one hundred, rush by, as if in great haste. Their arrival is announced by a most pitiable lamentation, calling out, "O idol, we, thy humble, spiritual followers, put our trust in thee and burn incense." When they have gathered around the altar, all kneeling, the leader beats a sonorous gong as a signal to the idol, and chants in a lamenting tone: "Saviour of the unseen world, save all the people from their sins and suffering." After which all the others repeat it three times in a similar pleading tone. A priest standing by the altar pronounces a benediction upon them as they depart, for which, of course, he expects a few cash. In like manner they make a complete round of all the temples, shrines, and altars, over eighty in number. We can only mention a few.

Passing up a romantic valley other temples for lodging pilgrims are seen, and after an ancient moss-covered bridge is crossed we are called upon to pass through the First Gate to Heaven. As might be expected, it is a very plain, humble structure. On the lintels of the door an inscription reads: "Those who have arrived here are to be considered as not outsiders, but as one of us." The chief idol here is *Ling Kwang*. He was originally a traveling mendicant during the *Sung* Dynasty, and became the first in rank of the twenty-six supernatural soldiers under the control of the "Jade Emperor." Above his head is a searching inscription: "*Shin wen er hsing*," (ask your own heart). He is believed to be a subtle and wide-awake god, having a third eye in the middle of his forehead. In another inscription it says of him: "With one stroke he awakens all men, and with three eyes he overlooks all under heaven. He helps people to gain their object in life with speed; his aid is as quick as thought."

Having passed through the first heavenly gate, the road leads through a dense bamboo grove, within which is located the Temple of Refreshing Dew, a name doubtless suggested by the rich, beautiful foliage with which it is surrounded. This monastery is gorgeously decorated and covered with imperial tiles of various colors; it is also honored by a tablet from *Kang Hsi*, and contains images of the Three Pure Ones, the Eighteen Lohan, the Goddess of Mercy, and *Ti Tsang Wang*. When I visited it a great Buddhist rally had just been held, at which two hundred priests had taken orders. A few more *li* up the mountain side the pilgrim enters the Second Heavenly Gate. This too is a small, unpretentious building. The road leads through the temple. In the passage is also a *Lingkwang*, and in a dark and sooty compartment is a Shakjamuni Buddha. At this stage it is claimed that all who accept Buddhism are moved upon and feel the affinity. Happiness is promised to all who come, and virtue administered to every pilgrim.

Beyond the second gate is the Dragon Pool Temple, an edifice of no special renown. A few more steps and the weary pilgrim reaches the Temple of the Southern Heavenly Gate, the half-way station. Such it is in reality up this mountain path from the first temple up to the seat of *Ti Tsang Wang*, and so it is in like manner the upward path of the Buddhist religion from earth to the ethereal heaven, called the beautiful and spiritual. Then the Temple of Ten Thousand Ages is seen, and the Third Heavenly Gate is reached. Here the pilgrim is welcomed under a large spacious passage, in which a god of riches and the king of the dragons appear in most gaudily painted colors. An attempt at beauty is also made in the inner hall, where the Goddess of Mercy with her servants is enshrined.

Following the winding road up over a ridge we saw an unusual scene. A *Lingkwang* of life-size was placed by the wayside, seated on a chair, with an incense burner in front of him, and a broken one to his left. The little basket by the burner betrayed the object of his appearance in public—a few cash being in it.

A little farther on, over a thickly wooded knoll, the illuminated glory of Buddha was seen—a restful valley nestled in the top of the mountain. Approaching from the north the Principal Gate of Heaven is seen on the opposite side of the valley. This gate leads to the Hall par excellence, also called the Mortal Body's Precious Hall. Here rest the mortal remains of *Ti Tsang Wang*;here his chief influence was exerted; and here he is worshiped by millions of devotees.

This building is square. On the east and west sides stand the Ten Rulers of Hades. In the southeast and south-west corners are two Police of Hades, one having an ox-head and the other a horse-head. In the northeast corners the civil and military judges preside. In the center of this hall stands a square altar. Upon this altar a pagoda is erected, which reaches up through the roof of the building. On the south side of this altar are five images of *Ti Tsang Wang* and two servants; on the north four images of *Ti Tsang* and six servants; on the west two images of *Ti Tsang*, one Laughing Buddha, and seven servants; on the east two images of *Ti Tsang* and two servants.

In front of this building and on the outside are these short inscriptions relating to *Ti Tsang:* "The moral body pagoda;" "These ornaments are precious ornaments;" "He confers grace upon all persons;" "His blessing extends over all Asia;" "His divine power has been shown in all directions;" "His intelligence in Buddhism was innate;" "The lucky wheel of fate stopped here forever;" "All people receive his divine favor;" "His kindness extends to all living beings;" "The divine clouds shelter all;" "This is the most divine place between ten thousand hills."

On the north, outside this building, are five large iron incense holders. The two outer ones have horns and are five feet high and four feet in diameter; the two inner are four feet high, and in the corner stands an incense holder with a cover like a Chinese umbrella. Upon this vessel pilgrims were rubbing cash, until they became bright. The belief is that children who wear these cash around their necks will become heroes. It is called the heavenly vessel. Around these incense burners is the rallying place of all the pilgrims. Here they kneel under the open sky and worship*Ti Tsang*, burning great quantities of sandal wood.

Beyond this chief temple is the Hall of Transmigration. At the entrance of the south door stand two black and red, ferocious-looking guards. At the north door two yamên runners with ox and horse heads. In the northeast corner is a scribe, who keeps the accounts of men's lives; by his side a fierce, black image, holding a banner upon which is written:

"Discriminate between good and evil." In the northwest corner stands a military officer holding a sword, upon which is inscribed: "The urgent warrant will seize quickly," and in the southeast corner stands two devil-faced, gaudily dressed yamên runners. The duty of these guards is to keep off all wicked worshipers from *Ti Tsang Wang*. On the east and on the west sides of this hall the ten divisions of Hades are represented, each presided over by one of the ten kings. They are holding court and mete out indescribably cruel punishments upon all criminals. Above these courts, from the first to the tenth, are the following mottoes: "Lay thy hand upon thy breast and examine thyself." "Here goodness and wickedness must be distinguished." "Who has concealed anything from me?" "Who has ever been forgiven by met" "You knew this condition before you came." "Why do you have so much trouble?" "There is no place for repentance here." "You can not do any better now." "The shore is just behind you," *i.e.*, when you could have reformed. A volume could be written upon these ten courts. That no beneficent influence has resulted from these terrible exhibitions of punishment, or is affecting the people at present, is apparent by the callousness of the worshipers when passing by these awful scenes of the future state. The reason no doubt is because these punishments rest upon a false basis.

In the middle of the ninth Chinese month, when the autumn moon is full, the visitor, standing on any of the surrounding peaks, can look down upon a scene both beautiful and pitiable. All that the creative hand has done is magnificent and grand, and the buoyant air makes one feel as if the elixir of life had indeed entered one's veins; but. that mottled throng down among the temples, the ceaseless thrilling echoes of the thousands of worshipers hastening from altar to altar, calling to the gods; the banging of boisterous blunderbusses and fire crackers, the din and noise of merchandise, and the pleading and moaning of countless beggars, create an impression far from what the valley is claimed to be—a fairy land. On our last visit we saw in one day no less than seven thousand pilgrims passing these altars. The great majority of them were young and middle aged men, but often, too, very old women were seen laboring under great difficulty to reach these heights. What weary travelers walk these roads! Occasionally a well-dressed young man is seen, wearing a tinseled head-band in apology for his mother, who has found it impossible to go herself. She will wear this band after death and appear with it before *Ti Tsang* in the next world. Recognizing the tinseled band, he will take it for granted that she worshiped at his altar in this life. Even gods must be deceived.

But nowhere can more misery be found in this or any land than is congregated here in these mountains during the pilgrimage season. It has become the harvest time of the maimed, halt, and decrepit of every description from the adjoining country far and near. They have united into an organized band, have built booths along the road from the foot of the hills to the highest temple ground. Every one expects, and some impertinently demand recognition from each pilgrim in the shape of small cash. Brokers among them collect these small coins and return to the foot of the mountain, where they sell one thousand of the small cash for four hundred large ones to the newcomers.

The last devotional act of the pilgrim is performed when they re-embark for home. On the bank of the river they kneel down, facing the mountain, and once more in their weird tones they chant their petition to *Ti Tsang Wang*—Earth's Hidden King.

* * *

POOT'OO: CHINA'S SACRED ISLAND

This islet is one of the Chusan Archipelago, situated in the 30° North latitude and 122° 25′ East longitude. This archipelago includes over one hundred islands and takes its name from Chusan, the longest of the group, which is over twenty miles long and six to ten miles wide. Poot'oo lies a little over one mile east of the Whang Head, is irregular and curiously shaped, about four miles long and very narrow at some places. The ceaseless march of the ocean tide has washed away all arable soil until the bare granite rocks along the shore resisted their incursion, except in a few small sheltered bays, where sandy beaches offer delightful sea bathing. Above the jaggy beach, however, enough soil remains, covering the coarse granite, to produce a rich foliage to the summit. The view from Lookout House, the highest point, about 970 feet above the sea-level, athwart the ridges and down the valleys upon the bestudded main, is highly picturesque and extremely delightful.

On approaching the island the visitor is landed at the southern point upon a well-built stone jetty, from which a broad, well-paved road leads to the three main temples. The front is called Universal Salvation Temple. The rear is called Rain-Producing Temple, and there is another called Wisdom's Salvation Temple. Smaller roads branch off in all directions, leading to grottoes, temples, and shrines of all sizes and shapes. As of ancient Judah, it can indeed be said that "in the high places and on the hills and under every green tree" the gods are worshiped, and even in the rocks images are hewn. These walks are lined with large, shady trees, and aromatic shrubs diffuse the air with a pleasant fragrance.

The numerous inscriptions chiseled into rocks along the pathways are an evidence that the whole island is devoted to the propagation of the doctrine of the Goddess of Mercy. Here are a few specimens:

"With a reverent heart take a look. The Law of Buddha has sacred affinity. The Goddess regards all men with kindness. Even the stupid stones bow their heads. The Buddhist kingdom together ascends. Ascend and enter the region of formlessness. Ascend and behold the Bodhi—supreme wisdom. Illusory light rises in the East."

On a stone tablet set on the top of a hill are the following inscriptions:

"There is a sacred island on the sea. Over the sea there is a Buddhist kingdom. I put my trust in Amida Buddha. There is another world."

On the opposite side of the tablet is the following:

"Only virtue is original. There is also Wisdom's Salvation Temple. With reverence be cautious not to kill living creatures. Do not pour hot water upon the ground, lest living creatures be injured, and when walking, be careful not to step on anything living. Such is the heart of the Great Conveyance and of Supreme Wisdom, necessary to the enlightenment of Buddha. With uprightness of heart cultivate the body. Be most careful to guard against avariciousness. If these worldly desires are not entirely exhausted the sacred fruit is not complete. Let all under heaven give reverence, and ascend to the other shore. Read good books and speak good words. Do good deeds and be good men. Imitate good examples and retain a good heart. Read the Buddhist ritual and worship Buddha. Abstain from meat with a reverent purpose, and do good in abundance."

In a conspicuous place the following exhortation is inscribed:

"In a Buddhist Classic there is a law in which it is said, 'kill not.' In explanation of this observe: above there are all kinds of gods, holy men, teachers, monks, and parents; below there are four footed beasts, birds, wriggling worms, and small insects. All that has life should not intentionally be killed. In this classic it is also said: `The winter months breed lice, take them and put them into a bamboo joint, keep them warm with cotton, and give them oily food to eat lest they might freeze or starve.' Such is the doctrine of the Goddess of Mercy."

Following the path that leads eastward along the shore until the northeastern point is reached, a place of unusual interest is seen, called the *Fan Yin Tung* (Buddhist Echo Cave). In an almost perpendicular rock, over one hundred feet high, is a wide cleft extending back into the rock so far that the end can not be seen from the temple bridge built across the chasm about midway. When the waves dash. against the rocks at the bottom of this cliff, and the sun shines upon the spray and mist rising up into the clefted rock, a natural phenomenon appears, the colors similar to those produced by the sun's rays falling upon raindrops are seen. The devout Buddhists firmly believe that this natural appearance is a living Buddha. A priest said to us, "Only believers can see him." When visiting this place, two elderly women were worshiping upon this temple bridge. After making their prostrations facing the cleft, the old Taoist recluse in charge of this shrine directed their attention to the place where the Buddha would appear. They looked, and they looked, until nature was in their favor, and the sun broke through the clouds, and they saw their heart's desire—the Living Buddha —a fraction of the rainbow.

Because of this belief, in the early days of Buddhism, many devout believers cast themselves down over this precipice in the hope of thus attaining to Buddhahood. A magistrate of *Tinghai*, hearing of this, wrote a proclamation in which he exhorted the people not to act so foolishly. He assured them that all who cast themselves down over this precipice would not be protected by the goddess, and would obtain no benefit; but the people did as before. Then he wrote a book and called the attention of the public to the moral nature of this abuse. "They who injure their bodies," said he, "injure their parents, and can not be considered filial. Think of the harm to yourselves in throwing your bodies among those rocks, where they will be dashed in pieces by the waves and eaten by the fish. And think of your families, your fathers and mothers, your wives and children. You came here to worship the Goddess of Mercy, to obtain blessings for your home, and word comes that you have destroyed yourselves. What grief! What sorrow! The goddess does not want such offerings; she herself will be distressed. If any want to sacrifice their lives, let them do so upon the altar of their country, and all will know that some good has come to the world through their devotion. As your official, I consider it my duty thus to exhort you." This book had its desired effect.

Since then the government of this island has become independent of civil jurisdiction, being ruled by the abbots of the chief monasteries. The abbot of the Rear Temple seems to possess more than ordinary business ability, judging from the clean and orderly condition of this temple. He is said to be an ex-compradore of Ningpo and Shanghai. Their mode of government is much the same as the civil government. A priest at *Futing-san* told us that even the power of administering capital punishment was in their hands, but the island being the home of supernatural beings and so many gods spreading spiritual light, corporeal punishment was seldom required. This priest had an exalted opinion of the deep spiritual enlightenment of their fraternity. When he was asked whether the older priests admonished and exhorted the younger, he indignantly replied: "How can we be exhorted when we have once comprehended the doctrine!"

The next point of interest was the many miraculous legends told and believed by the people. The most of them have the coloring of historical authenticity, giving names and dates, but they lack in credibility. The oldest reach back as far as the T'ang Dynasty. We can only give a few:

The Emperor *Tai Ho*, who reigned near the middle of the ninth century, 827–836, was said to be passionately fond of *holi*, a species of clam. So the people of these isles yearly sent him all they could gather, even greatly overtaxing themselves for their Emperor. One day, as he was eating of his favorite dish, he found one with a hard shell which he could not open. He cleaved it with a knife and found within an image of the Goddess of Mercy. When he saw this he was frightened, and commanded that the image be incased in a sandalwood casket, overlaid with gold, and placed in the Imperial palace. Then he asked a priest named Wei Chen what this meant. The priest replied: "The goddess desires to open thy heart that thou mayest be temperate in all thy desires." Whereupon the Emperor no longer pressed the people to send him clams, and issued a proclamation that an image of the goddess should be placed in every temple.

Another story is told of the same period, that a priest burned his fingers to show his zeal for the goddess. When his fingers were almost burned off, she appeared to him and comforted him for his faith and devotion.

During the Posterior Liang Dynasty, in the reign of *Chen Ming*, 915, a Japanese priest named *Hui Ngo* brought an image of the goddess from *Wutai*, and was going with it to his home in Japan. When his boat came into the sea of the water lilies, near Poot'oo, it was in danger of being wrecked. So the priest prayed to her and said: "If it is thy pleasure to go with me to my country, I am willing to go with thee wheresoever thou wilt have me to go." After this the boat smoothly glided on its way, and soon came to the landing in front of the *Chao Yin Tung*, the Tidal Echo Cave. Below the cave lived a man named *Chang*, who gave his own house to the priest for a temple, and the temple was named "The Unwilling to Depart Goddess of Mercy Temple." It is not now in existence.

In the *Sung* Dynasty, the Emperor *Yüan Fêng*, in the third year of his reign, 1081, sent a minister named Wang *Shun-fung* to Corea. On his way he encountered a terrific storm, and a large sea monster came to overthrow his boat. Terrified he prayed; as he was looking intently toward the cave in the distant island he saw the goddess coming out of the cave, dressed in gold apparel and adorned with rich jewels. Immediately the monster left him and the sea was calm. He returned in peace and informed the Emperor of his experience, who ordered the place to be called the "Precious Declivity."

In the *Sung* Dynasty, the Emperor *Ch'ung Ning*, in the year 1102, sent two Ministers of State to Corea, named *Lin* and *Wu*. On their return they passed through the Chusan Archipelago. When sailing among the islands, a dense darkness came upon them, and for four days they did not see either sun or moon. In their distress they remembered the goodness of Goddess of Mercy on Poot'oo, and they worshiped, when suddenly the surface of the sea was illuminated with a brilliant light, and they proceeded rejoicingly on their way, and soon saw *Chao pao san*, the Hill of Precious Beckoning, near *Ningpo*, from which they went to the mainland.

During the Southern *Sung* Dynasty, the Emperor *Lung Hsin*, in the year 1163, dreamed that he was in Poot'oo, where he saw many wonderful signs of the greatness of the Goddess of Mercy. After this he composed an ode in which he praised her greatness, ascribing to her the ability of accommodating all persons, and of being able to reveal all mysterious doctrine. And he called upon all his subjects to do her homage, for no other god was equal to her in wisdom and goodness, promising to all men what they prayed for. "Her mystery man can never understand," he exclaimed.

In connection with the image which refused to go with *Hui Ngo* to Japan another story is told. This image was continually working miracles, so that the people of other places and cities also desired her patronage. Whereupon a certain priest came to Poot'oo, bought a good piece of sandalwood, entered the temple and closed the door. He worked for one month, carving a facsimile image and disappeared with his newly-carved goddess. In the Southern *Sung* Dynasty, during the first year of the Emperor *Chia Ting*, 1208, the image which the priest had carved lost a finger. When the priest in whose temple the image was at this time saw the mutilated limb he was terrified. In his agony, as he was looking towards Poot'oo, behold a flower came floating along from the shore below the cave, bearing the missing member. This image is now in *Chao ying tung*.

In the Southern Sung Dynasty the Emperor *Hsien Shun*, during the year 1265, had a high official named *Fan*, who had diseased eyes. He sent his son to Poot'oo to pray. The prince brought some water from a spring below *Chao ying tung*, with which the statesman washed himself and was cured. Again the Emperor sent his son to offer thanks to the goddess. Having discharged his obligations, he was sitting by the cave when the goddess revealed herself in a cloud of smoke in which she was veiled. When the young prince came to another cave he saw the servants of the goddess standing face to face by his side. They were dressed in white garments with crowns upon their heads, both looking towards him as if they wished to speak.

The following two legends are assigned to the *Yuan* Dynasty. In the thirteenth year of the emperor *Chih Yüan*, A.D. 1264, a high official was sent with a Lama priest to the Southern isles to restore order. Upon his arrival he expected to see the goddess as others had seen her, and when no vision appeared he was angry and shot an arrow into the cave and took his departure. On his homeward way he was surrounded by water lilies so dense that all progress was impossible. Fear fell upon the warrior, and he repented of his unbelief and hasty action. Returning to the cave, the goddess revealed herself to him, dressed in white and attended by her maid servant; whereupon he ordered a temple to be built over the cave.

In the fifth year of the Emperor *Ta Tê*, A.D. 1302, a man of literary distinction, named *Chu san fung san*, was sent to worship on the *Li* Mountain. On his way hither he passed *Chao ying tong*, where the goddess appeared to him, wearing a crown and necklace of precious jewels, holding in one hand a willow branch and in the other a blue glass cup. During her visible presence the cave was filled with light.

The *Ming* Dynasty was specially fruitful in producing interesting legends. In the twenty-second year of the Emperor *Yung Loh*, A.D. 1424, on the tenth month and nineteenth day, the goddess revealed herself in the Tidal Echo Cave, dressed in white and attended by a dragon king and his daughter, who were accompanied by a host of servants. From ten to eleven o'clock they opened wide their eyes and made a sorrowful countenance. By twelve o'clock their countenance had changed to purple and they stood facing the wall. At six P. M. they were seen at the Archer's Cave, and a god dressed in white, wearing a golden crown, was seen sitting in the midst of the sun, and *Bodhisattoa* was standing under the sun, while the *Lohans* were walking upon the sea. On the following morning at six o'clock a god with a purple gold body was seen in the cave.

In the second year of *Hsüan Tê*, A.D. 1428, and the fourth month, *Yen Loh Wang*, the Ruler of Purgatory (the Chinese Pluto), with his two daughters were seen in the cave.

In the year 1575, the second year of *Wan Li*, a priest, named *Pieh Chuan*, was on his way upon the sea to Poot'oo, when he saw in the sky a goddess dressed in white, flying westward, while another priest, named *Tsai Fung*, saw her passing over Chêntu, the capital of Szechuan.

Six years after this a priest named *Ta Szyung* came to this island and found the conditions very favorable. The desire of his heart was to build a temple, and he came to the cave where the Goddess of Mercy dwelt, and prayed, "If I should build a temple here, give me a sign." In the evening of the same day he saw the tide bring in a large bamboo. "There," said he, "is the sign," and he built a Tidal Wave Temple.

In the fourteenth year of the Emperor *Wan Li* there lived a Hangchow priest at Poot'oo, whose mother was a most ardent worshiper of the gods. She was also a vegetarian, and went about collecting subscriptions for temples. One day as she was soliciting gifts an idol was given her. The neck of this idol was of gold, and she gave it to her son, the priest. When he saw that the neck was of gold, he coveted the valuable metal and engaged a workman to peel it off; but no sooner had the workman begun his sacrilegious work than he dropped dead. Soon after this the mother came in her wanderings to Poot'oo. When the priestly son saw her he began to curse and said, "The one who injured me has come," and he dealt her a slap with the hand. After rushing over the hills in his madness he cut his throat, but before dying he said to his colleagues, "Follow not my example, for if you do, you will come to the world of misery."

Four years after this, on the tenth month, the priests of the island had an altercation which they could not settle among themselves. So the case was submitted to the magistrate at *Tinghai*. When he came to the island to settle their trouble, his faith in Buddhism and the pretensions of the priests was thoroughly shaken, and he took a set of the standard Classics of the Lotus School and burned them and made the priests walk over the ashes. That night he had a dream. In his dream he saw a sacred man coming to him, who explained to him the meaning of the Buddhist doctrine, and then said: "Heretofore you believed this doctrine, and now you persecute it; for this you can not obtain forgiveness. In the spiritual world you will have to become a herdsman." Hearing this he was deeply grieved, and he begged the idol to forgive him; also a priest named *Ta Chih* prayed to the idol for him. Soon after this he had another dream, when Buddha appeared to him and said, "Forgiveness may be granted if you replace one hundred volumes for every volume you have destroyed." With this the magistrate gladly complied and returned to his island city, where he had one hundred sets of the Classics printed and sent to Poot'oo.

The following are of the Ch'ing Dynasty. In the reign of the Emperor *Shun Chih*, about the middle of the seventeenth century, there was a pirate chief named *Nü Chin*, who consulted with a Japanese priest to take to Japan all the Buddhist books which the Emperors of the Ming Dynasty had presented to the monasteries of Poot'oo. When a priest named *Chao Chung* came with several hundred priests and begged the chief not to rob them of their books, he became angry and said, "If you want these books, you will have to descend into the depth of the sea and confer with the dragon king in his yamên," and forthwith he set sail for Japan. When far out upon the ocean, behold, a great fish prevented his boat from going any further. All skill was baffled, and he was in great despair. He repented of his evil and returned, when a fair wind brought him to Poot'oo in less than half a day. The priests all rejoiced and replaced the books in their respective libraries.

In the third year of *K'ang Hi*, the first month and the first day, the priests at Poot'oo saw a crescent-shaped white light resting with one end upon the top of a temple and with the other upon a hill not far distant. And they saw the goddess dressed in white walking from the temple over this arch to the adjoining hilltop. From this they concluded that evil would soon befall this temple. One year later foreigners came to this island and took all the idols, banners and curtains, and carried them off to Japan, where they were sold for Tls. 200,000. On their return to their native land their ship was destroyed by fire, and they all perished in the sea. The same year pirates made a raid upon the island, burning the temple and smelting the copper images. When they came to the Rear Temple, the robbers, eager to plunder, pressed into the inner parts, while the front was set on fire, and thus they perished by their own hands.

Some years after this, in the Spring, when the tide was high, a fleet of gunboats passed this island. The sailors went ashore and found the temple all destroyed and the place clean, as if no one had lived there. Upon their return they were pursued by numerous poisonous snakes, also many vicious dogs came out of the jungle. Many of the sailors were bitten and perished. All because the god had been ill-treated and the temple destroyed. During the latter part of this year the *Ningpo* admiral came to this island with his ship, and his soldiers found a bronze image which they took with them. When they were on their way, the ship became so heavy by the weight of the god that it began to sink. The admiral, suspecting that something was wrong, searched the ship. When he found the god he ordered him to be returned, and his ship regained its buoyancy.

In the twenty-first year of *K'ang Hi* a literary man named *Kao-Sz-chi* wrote the book of Rewards and Punishments, a kind of commentary of modern Taoism. His wife embraced the Buddhist religion and also read the *Chin Kang Chin*, the *Diamond Sütra*—the *Sutra of the Pradjna*, which is able to crush diamonds. In the fourth month of this year she became ill and knew herself that she could not recover from her illness. On the twenty-seventh day she became unconscious, but soon revived, when she repeated the Sutra and said: "If I can but re-establish a temple on Poot'oo Isle, I shall live two years more." But on the fourth day of the following month she died. Before her death her husband requested of her a petition. "If in the Kingdom of Shades you really do not suffer agony, I want you to let me know through a dream." She nodded with her head and departed. On the second month and the seventh day of the following year her eldest son had a dream. In this dream he met seven or eight aged priests walking in a forest. Said one of them to him: "Would you like to see your mother? If so, come into my sleeve, and I will take you to Poot'oo." Immediately they were on the island, where he saw many large temples, and in one, named the All Pervading Majestic Temple, a number of priests were chanting prayers. One of their number said: "If any one will repair this temple his merit will last through all generations, and neither in this life nor in the life to come will he be exposed to the turning of the wheel and subject to transmigration, but in heaven and among men he will have all kinds of blessings." This promise spoken, he saw his mother dressed in a yellow and pink coat and a white skirt, folding her hands, as if in prayer. She said: "The doctrine of the *Sûtra* fills all space, its grace reaches to all, and all men under heaven should honor it. Through the help of Buddha I have learned all this doctrine and have not forgotten a word. My heart is fixed. I have attained supreme intelligence; in all eternity I can never be changed. Tell this to your father, that he may not be concerned about me." While thus dreaming, he heard the rushing of water, and awoke. When the father heard this, he said: "This, my son, is honest, and would not tell a lie, hence I will write what he has seen and heard in his dream upon my garments." On the first day of the tenth month he burned these clothes in sacrifice, that his departed wife might see his devotion.

In the year 1691, the twenty-ninth year of *K'ang Hsi*, the spirit of miracles seems to have departed from Poot'oo. It was during this year that the *Chentai* (General) of *Tinghai* came to the Buddhist Echo Cave, where a god appeared to him. His face was red, eyebrows heavy, beard long, eyes white and brilliant, and his nose a mere white speck. He wore a crown which shone like a flame of fire, and his garments were yellow and black with a broad collar, all loose and flowing, with one arm bare. Upon his head stood a little bare-footed *Bodhisattoa*. When the general made prostrations before this goddess, he (the god) disappeared. The god mentioned here is a metamorphosis of the Goddess of Mercy, which may be male or female.

No description of Poot'oo could give satisfaction without trying to describe at least one of the many temples. All sizes included, there are doubtless nearly one hundred. We choose the Universal Salvation Temple, also called the Front Temple from its location, being the first large temple the pilgrims see when coming from the landing. Like most temples in China, it faces the south, lying in a quiet valley, basking in the noonday sun. All that nature and Chinese art have done has rendered it quite an enchanting scene. The main buildings are covered with green and yellow tiles, indicative of imperial distinction, and are kept in a remarkably good condition. But an indisputable characteristic of the Chinese is here also clearly seen. The buildings, which cost them nothing, are very much neglected and in a dilapidated condition. The two pavilions housing the Imperial Tablets of *K'ang Hsi*, standing in the most prominent places of the two largest temples, are not only neglected, but subjected to a most sacrilegious use, having practically become the rendezvous of barbers and beggars, with all their belongings. We would not attempt a minute description of this temple, but only give an outline of its arrangement. We may say here that the Chinese seem to have great liberty in arranging the position of certain gods. It would be difficult to find two temples exactly alike.

Approaching this monastery, the visitors enter a shady avenue, where stands a stone arch of considerable age. It much resembles the ordinary arches which perpetuate the memory of virtuous widows, of whom China seems to have had many. Upon this arch are Tibetan inscriptions, a conclusive proof that Lama priests from Tibet or Mongolia have made pilgrimages to these shrines. Between this arch and the temple ground proper is a large lotus pond, spanned by a moss and ivy covered bridge. This pond teems with fish, and is the receptacle of much that might seem distasteful to these sacred vertebrate animals, but they leisurely seem to enjoy it. Beyond the bridge stands the pavilion, which contains three tablets. The center tablet is of a dark blue marble, fifteen feet high, five feet wide, and one foot thick. It stands upon a pedestal two and a half feet high. The inscription upon it was written by the Emperor *K'ang Hsi* in the forty-third year of his reign, the eleventh month and the fifteenth day. The following is a free translation:

"Studying the Buddhist religion, I learned that there were sacred mountains. One of these is located in the Southern sea. These books, however, did not give much accurate information. During the change of dynasties front the *Ming* to the *Tsing*, we were tossed in revolution all around; the island was overthrown by pirates, and all the temples were destroyed by fire. After the twenty-second year of my reign peace and order were established. Priests returned from the mainland, looked up the old foundations, cleaned away the weeds and debris, and began to build anew. When I came to the *Cheh-kiang* province, I sent an official to reverentially rebuild the temples and worship, and I myself prayed, saying, "May the temples forever keep (the sea) in subjection." Having prayed thus, I also sent gold for the repairs of the temples. The corridors and windows I ordered to be embellished with red and green, and all the beams beautified. Every beam and every stone was given by the Government, and all the artisans and laborers were paid by the Government. This I did for my parents above me and my people below me.

"When I was young I read historical classics, in which I learned how to cultivate the body, the home, and to execute government affairs in important matters. At that time I had no leisure to read the litanies, and they seemed to me as empty shadows, for I did not know the meaning of the mysterious Buddhist doctrine. Now I have learned that heaven is the greatest good, and the goddess rests upon this good, hence there can be no difference between the two. Heaven delights to create and give life to all creatures. The goddess is of great mercy in saving all that has life, hence she is not second to heaven. I have prayed for the Government to be active on behalf of the people for over forty years. And although there is no more rebellion in the land, yet the people are not fully at rest. The people are not yet pure-minded, and the ignorant are easily moved upon. In times of floods, droughts, and failure of crops, unseemly depredations are committed. These things never left me, even though I was asleep. I trusted to the great mercy of the goddess. I prayed that clouds of mercy, showers of rain, sweet dew and favorable winds might come to give a plentiful year to the people, and benevolence and longevity might prevail. If thus the goddess could save the nation and gain great merit, great good would also come to the world. This is my honest wish. Therefore I have written it to be cut into a stone tablet, that men may read it for all ages to come."

One month later he wrote an eulogy on the Goddess of Mercy, which is inscribed on a tablet at the Rear Temple. In this he says:

"I have heard that the goddess is all wise and has a wonderful appearance, that she is the origin of all truth and has opened a way to experience, giving evidence of the wisdom of the gods. Coming to this island, the moaning of the waves and the chanting of prayers can be heard, the deep purple temples can be seen, and peace can be had as expansive as the wide sea. But the goddess has other temples on Poot'oo besides this rain-producing temple. This renowned island has become a kingdom of the gods. It is like a ship of mercy upon the great sea; hills blue as the fleecy sky, and high as the heavens of Brahma; upon this lucky, clean place the waves dash up to bathe the sun. Its reputation stands as a pillar, supporting the sky. From its summits all places are connected.

"But when a time of trouble came, the priests scattered, as the morning clouds and the ashes of the burning temples were driven by the wind. Now, since happiness and peace have been restored, and the waves of the troubled sea cease to roar, the priests longingly looking to this place have returned. Special imperial grants were bestowed to aid in rebuilding the temples, to accomplish which it took much time, but they wearied not. When laying the foundations they beat a drum to stimulate the workmen to activity. In building, they did not erect pearl halls with precious doors, but used the huge turtle for the base of the pillars and the *iguana* for beams, that they might know how to cross the stream of error, that leads astray. The precious image of the goddess they seated in a natural colored lotus flower.

Having furnished all, they placed a shining wheel of the Buddhist law on the building, and planted the purple bamboo in the forests. When majestic images were arranged in all the halls, then clouds of mercy spread over the earth, and the Emperor gave the temple the beautiful name *Fa Yü* (Law of Rain). Heaven can boast of protecting the doctrine of the gods; for they are established high, like a beacon cloud, and ships will come from all kingdoms to do them homage. Thus all can cross the stream upon the raft (Buddhist doctrine), and ascend the other shore.

"Now, since the great temple is completed, I pray that my mother may be blessed with long life; for this I borrow mercy. May my people also obtain many blessings! And may benevolence and longevity abound, and all under the brightness of the moon be filled with mercy. Let this be carved into a good stone tablet that all who come may read it."

Leaving this pavilion, we see to the right and left two little three-storied kiosks, half hidden in the shady grove. In the one to the right is an image of a patron saint, in the other an image of *Ti Tien Wang*, the ruler of the Ten Halls in Hades. The first temple building we enter might be denominated the Protection Hall. Like most regularly constituted Chinese monasteries, it contains images of the four great kings, standing near the east and west wall, facing each other, *Milehfu* sitting on an elevated altar in the middle of the hall, and *Weit'o* standing on the opposite side looking north.

These four kings were brothers, and lived about 1100 B.C. They are supposed to be protectors of *Milehfu*, governing the center of the universe. Their faces are red, green, white, and black. The one with the red face holds an umbrella in his hand, with which he overshadows the universe. If he opens it, all is turned into chaos, sun and moon refuse to shine, and darkness covers the earth. The one with a green face holds a zither in his hands, and when he strikes upon it fire and wind come forth. The black faced one looks ferocious, and is supposed to be the eldest of the brothers. He holds in his hands an animal, which he has charmed into submission to do his will. The white faced one seems less ferocious, and holds in his hands a sword. When he brandishes it myriads of arrows fly in all directions, destroying all that would injure Buddha and his doctrine.

Between the four kings, in the midst of the hall, sits, in a glass encasement, "Laughing Buddha," the Merciful One, who is expected to appear and open a new era about three thousand years hence, when he will take the place of Shakyamuni. He is the Coming Buddha, the Messiah. His friendly mien invites the worshipers to come to him with all petitions. His devotees are chiefly women, who beseech him that in the next world they may be born men and not despised women.

Weit'o has a sword in hand, entwined with a gilt dragon. His stately bearing well becomes him as celestial protector. All the gods in this hall are still in the "wheel of the metempsychosis."

The space between this hall and the principal building is about one hundred feet wide. In it stand two miniature iron pagodas and a huge bronze incense-burner. This building is one hundred and fifty feet long and seventy feet wide. Two rows of pillars, six in each row, support the imperial yellow and green tile roof. Upon entering this lofty hall we see in the corner to the right an image of the celestial protector, Weit'o; to the left, a large drum over ten feet in diameter, and in the midst of the hall the main altar. At the front door, and over the altar, are curtains beautifully embroidered after patterns evidently very antique. In front of the altar are two brass incense holders, urns for flowers, and eight candle supporters. The tinsel and ornamental accessories upon the altar and images can not be described. Instead of the Three Pure Ones, *Shakyamuni*, *Kaskiapa*, and *Ananda*, upon the raised altar, as in the other Buddhist monasteries, there are here four images of the Goddess of Mercy. The temple being dedicated to this deity, she naturally occupies the most conspicuous place. The four stand in a row from the front of the altar to the rear. The first is the goddess standing. Directly back of her is a large gilt image sitting cross-legged upon a lotus flower. This image was sent from Tibet by the Emperor *K'ang Hsi*. She is beautifully draped with yellow silk, and wears the crown of the *Boddhisattoas*. The third stands in sacred reverence with a halo over her masculine head. Behind her is a monstrous male god, supposed to represent the ruler of the monastery, and who is denominated by *K'ang Hsi* as "The Savior of all Living Beings." All of these images have servants, except the male gods. To the right and to the left of this altar are images of *Kuo Hai* and *Sung Tsz*, that is, the goddess that crossed the sea, and the giver of children. In place of the Eighteen Lohans (disciples of Buddha) along the east and west walls, there are thirty-two metamorphoses of the goddess, the Lohans being relegated to side chapels.

In a little room adjoining this building a scene of the lower world is represented. A Lohan is sitting upon a one horned sacred ox, which is standing upon a jagged rock. Under the rock is a cave. In the cave an official is holding court. Before him a man is kneeling. The executioner is near with his ax. To the left stands *Weit'o*, and to the right *Ming Wang*, the ruler of Hades.

The third building, directly back of the main hall, is exceedingly plain. In the northeast corner is a drum, in the southwest hangs a bell, in the northwest stands a *Fu*, holding a little pagoda in his hand; on the center altar are the Three Buddhas, all in perfect keeping with the plainness of the whole building. Upstairs, in a hall almost void of furniture, sits a goddess hewn from a precious white stone, found in Southern Tibet, called by the priests *pehtsz yü*, or white jade. She is loosely draped with a red silk cape, and were it not for the cynical smile playing upon her face, she might be considered a rare beauty. This room also contains the library.

The fourth and last of the temple buildings is even more plainly furnished than the third, but it contains, beyond all doubt, sculpturally, the finest work of art we have seen in China. It is a small image of the goddess, about three feet long, of like material as the one in the library building, a white, watery-looking jade. The image is modestly lying down within a glass case, gracefully resting her head upon her right arm, and is covered with beautifully embroidered red silk—a speaking likeness of the goddess of beauty. Visitors should not fail to look into this hidden compartment.

Near the monastery is a small stone pagoda four stories high. It dates from the *Ming* Dynasty, being dedicated to the Emperor *Wang Li*, who reigned during the latter part of the sixteenth century, and it is probably the only structure on the island which survived the change of dynasties. The architecture is quite unlike modern Chinese, rather resembling Hindoo art. Twelve images are hewn in the four sides, which are supposed to have control of the elements of nature.

But what interested us more than the beautiful scenery, works of art, and strange legends was, in the first place, to observe the unmistakable keen sense of chastity in their idolatrous worship. Free from all vulgarism and meretricious ornaments, there was not even the suggestion of what might be offensive to the moral sense. However much we may pity the Chinese in their ignorance in worshiping the workmanship of their own hands, we can not help but admire their belief in the purity of their gods. The next point of supreme interest to us was the people who visit these shrines. To see something of their devotion we took passage from *Ningpo* on a regular pilgrims' sailing boat. Forty-odd persons crowded on to the little barque. The room was quite insufficient for that number to lie down, but for one day and two nights they patiently sat without a murmur or complaint. They had come from various provinces, *Chih* in the north, *Szechuan* in the far west, and a goodly number from *Hunan*. Stoic resignation and privation seemed impressed upon every face, not because of a lack of vitality, but on account of the vows they had taken.

The earliest mention made of this island reaches back to the *Liang* Dynasty, A.D. 502–556. During the *Sung* and *Yuan* Dynasties reports are recorded of building and repairing temples. During the present dynasty special efforts have been put forth, both to beautify the temples and to increase their number, in the belief that the gods may spread great spiritual light. Let all who believe that the worship of the Goddess of Mercy is on the decline visit Poot'oo.

THE WHITE DEER GROTTO UNIVERSITY

On the southeastern slope of the Lü Mountains, in a little valley at the junction of the two rippling brooks, one flowing east and the other south, stands one of the oldest universities of which we have any knowledge; Salerno, the oldest European university, not excepted. With *Wu lao fung*, "five old peaks," standing like parapets on a rampart for the background, and the *Poyang* Lake winding up the valley beyond the lower undulating hills, it can claim a situation at once attractive and grand. The grotto shows no sign of ever having been a natural cave, being dug into the hillside and arched over with masonry work. It contains the image of a deer hewn out of stone placed there by *Ho Tsing* in the fourteenth century. According to historical statements it was the studio and home of the illustrious poet *Li P'u*, who flourished during the T'ang Dynasty in the reign of *Chêu Yuan*, the latter part of the ninth century. He and his elder brother *Shê*made this cave their private home. *P'u* had a tame white deer which always accompanied his master in his walks over the hills; and thus he was called *peh lu sien-seng*, "the white deer gentleman," and his residence the cave of the white deer.

During the reign of *Pao Li*, A.D. 825, P'u was promoted to be sub-prefect at *Kiang-chou*, now *Kiukiang*. To perpetuate the memory of his old sequestered-home, he built a kiosk over it, from which time the cave of the white deer became famous. At the end of the *T'ang* Dynasty, when there was great confusion in the Empire, the scholars from far and near assembled in the cave for research. During these troublous times when five successive dynasties rose and fell in two years (A.D. 805–807), a school was opened here, fields bought, buildings erected, and students gathered. The scholarly *Si Shan-tao*, who had thoroughly mastered the Nine Canonical Books, was chosen from the Imperial Academy of Learning to be the president of this institution, and it was called the Government School of the *Lü* Mountains. The institution soon increased in number and influence, and many of the early students became famous in public life. In the year A.D. 960, the institution was raised to the grade of a university, and stood equal in rank and influence with *Chi Yong*, *Shih Ku*, and *Yo Lao* Universities. The attendance of over one hundred students is mentioned at this time.

In the cyclical year *Ting-cheo* (A.D. 976), during the reign of the two Emperors, *T'ai Ping* and *Hsin Kuah*, the sub-prefect of *Kiukiang* begged the Emperor to present a set of the Nine Canonical Books of the Imperial

Academy of Learning Edition; this modest request was granted, and the books were sent to the cave by the government postal service.

The institution passed through various changes, more failures than successes following each other, until the cyclical year *Chi-hai* (1174) in the *Sung* Dynasty, and the reign of *Shun Hsi*, when the great philosopher and expositor of the Confucian Canon, *Chü Hui-ngan*, became the prefect of *Nang-k'angfu*. No sooner had he arrived than he ordered the official director of students to proceed at once to the cave and investigate conditions. Later he personally went there and carefully looked all the property over, whereupon he sent a message to the President of the Board of Works, informing him that the building might be repaired, and also reminded. him that the White Deer Grotto was a place where the former worthies had concealed themselves, and where students had been supported by previous dynasties, which favors should not be forgotten. In pathetic terms he appealed for help: "The buildings are falling down, and the prefect can not but take the burden of repairs upon himself; he has measured the place and calculated the expenses. Why should he do this? Because the name of this institution is recorded in the national annals."

After a lapse of time he sent another message to the Senior Secretary of the Board and to the Minister of State, begging to be appointed President of University, and to ask for some grants to be bestowed. If this could be done, he would be asking for no other favor in his life than to be allowed to study with his students until death. He also pointed out how much better this would be than to add glory to the heterodox priests who only burn incense and do nothing for the good of mankind nor for their food and clothing. And he further pleads that this institution receive recognition as well as *Yu Lao* University and a tablet be bestowed naming it the White Deer Grotto University, and that some parts of the classics be written upon this stone tablet by the Emperor *Kow Tsun* himself. It is well known that at this time the publication of the Classics of Confucius and Mencius was forbidden, in consequence of which education had also fallen to a low ebb, and all classes, the officials and common people, complained of this, feeling deeply chagrined. When *Chü Hui-ngan* was acting as Inspector of the State Department, he had an audience with the Emperor, and he again made a plea for more liberal education. Said he: "The Taoist and Buddhist temples are built everywhere. In the provincial capital there are more than one hundred, in every prefecture several tens, all well established publicly and privately, and yet there is desire for more; while there is only one school or college open in a prefecture, and small districts have none. Thus temples are prosperous, while schools and colleges are neglected; temples are numerous, while schools and colleges are few. Why this inequality? Is it not wrong that the civil authorities do not make this right, and that they even look upon any information with suspicion? If this continues the Grotto University will soon be despised by the common people, and not be kept open unless an Imperial Tablet be bestowed. Is it not the way by which your Majesty can praise your meritorious predecessors and do the scholars a favor? This I venture to beg at the hazard of my life." When the Emperor *Hsiao Tsung* heard this he granted the request; but the tablet did not assure the institution perpetual peace and blessings. At the end of the following dynasty—the Yuan—it was cast out into the brushwood during a commotion, and not until the sixth Emperor of the Ming Dynasty was it found and replaced.

History does not clearly state how long *Chü Fu-tsz* labored here, but hereditary legend claims that he spent the rest of his life in this institution, and was buried in the shady grove back of the college. The legend also pretends to know the origin of his superhuman wisdom. When he came to live at the grotto, a *hu li sien*, "fox-fairy," in the likeness of a young woman, came to live with him and serve him. She brought with her a pearl of great value, and insisted upon *Chü* swallowing it. After long persuasion he yielded to her entreaties, and the pearl became in him the fountain of wisdom, such as is not possessed by mortal man. Soon after this a *ka ma sien*," frog-fairy," also in the likeness of a young woman, came to dwell with him. But alas, the women did not live together long in peace. In an altercation the last arrival said, "And who are you but a fox?" Replied the other, "And what are you but a frog?" The next day the two fairies were missing, and a dead fox and a dead frog were found lying under the old bridge below the college. They were buried with due ceremony in the college grove, where a little stone marks their resting place, to the credulous belief of the student and admirer of *Chü Fu-tsz*.

A visit to this historical spot is least desirable in summer, when the students are here from all parts of the central provinces; during the Chinese New Year vacation it is a pleasure and profit. If a well-informed student can be secured as guide, much that seems to have no significance becomes interesting. Being thus provided, we left *Kiukiang* two days before the New Year, early in the morning, and arrived there late in the evening. A young man from *Fou lan hsien*, an unsophisticated youth, who had never seen a foreigner, was the only dweller within these walls. He had come here to spend the holidays, having heard that a man having done so the year before, worshiping Confucius on New Year's Day, took his degree at the next competitive examination. On the morning of the last day of the old year he walked five miles to *Nank'ang* to buy candles and incense, and that night and the following one he kept them burning upon all the altars in those labyrinthian rooms. He was radiant with hope.

If the present condition of this institution is a criterion of the state of the religion it stands for, then Confucianism is in a most hopeless and woeful plight. It is an institution without a recognized president or faculty, without a Board of Trustees, or even a janitor. The *tipao* was said to be in charge, but during three days' stay we did not see him. The director of the literary class of *Nank'ang*, having heard of our arrival, called upon us, but he did not assume any responsibility. The students seem to be a law unto themselves. They bring their own cooking utensils with them and build a little hearth for their private use; some unite in clubs. This explains the dilapidated condition of the buildings; whole sections of the wooden partition being broken out and used for fuel. Over some parts the roof is crushed in, and weeds are flourishing in the rooms. The memorial tablets, of which there are many, have sagged in all directions, and many have fallen down and are broken. It is now nothing more than a quiet place for students to hide away from the disturbances of home life, and so be better able to prosecute their private studies.

The Literary Assembly Hall is a large, substantial building, but entirely void of all furniture, there being not even a scroll to decorate the black walls.

But the learner here stands face to face with the eight virtues which are indelibly inscribed in large, bold characters upon stone slabs set in the walls:

Hsiao—Filial Piety.
Ti—Respect of younger brothers.
Chung—Loyalty.
Hsin—Sincerity.
Li—Propriety.
I—Righteousness.
Lien—Purity.
Chih—The feeling of shame.

On the inside of the back door of this hall is written an amplification by *Chü Fu-tsz* on the Five Relations as taught by Mencius; upon these *Chü* based his Rules of Order: "Between father and son there should be affection; between sovereign and minister righteousness; between husband and wife attention to their separate duties; and between old and young respect; and between friends fidelity." These are the instructions regarding the five relations, and the observing of them was also enforced by Yao and Shun, who appointed Su to be the minister of instruction and to teach carefully the relations of humanity. The students should learn them in the following regular order: Study them extensively, enquire about them accurately, reflect upon them carefully, discriminate clearly, and practice earnestly; such is the order of learning: study, enquire, reflect, and discriminate thoroughly by examining the principles. To practice earnestly refers to the cultivation of moral conduct, the managing of business affairs, and the making acquaintance with others.

"In regard to the cultivation of moral conduct, the most important thoughts are: let the words be sincere and truthful, the actions honorable and careful, the anger restrained, and the lust chastened. Reform and be good.

"In regard to the managing of business affairs two things are necessary, the action must be orderly without scheming to gain profit, and the doctrine must be thoroughly understood without counting too much the toil in getting the mastery of it.

"In regard to making acquaintance with others, the important part to be remembered is: not to do to others as you would not wish others to do to you; and when you do not realize what you need, then turn inward and examine yourselves in every point.

"We learn that ancient worthies instructed men to investigate the principles of righteousness and to cultivate a moral conduct by which they might influence others. They did not merely wish men to commit to memory, write compositions, by which they might gain fame,' vaunt themselves and gain profit. But the student of the present time (the days of *Chü Fu-tsz*) do not follow the methods of instruction as taught by the ancient worthies. Let the students who have earnest thoughts give due attention, inquire and discriminate. If one knows what ought to be done and forces himself to do it, will he not eventually know intuitively what his duties are without rules of order? The rules which have been established are for students of a lower grade than those of ancient times. And as the methods of the present day students do not agree with those of the ancients, I do not put the rules of ancient methods in this hall, but on the lintels of the door. Those important subjects which the ancient worthies taught, I myself will investigate and follow with all the students, and we will force ourselves to practice them. Moreover, we will endeavor to be even more rigid and careful what we are thinking, speaking, and doing, than these subjects require of us. And those who are inclined to abrogate and neglect these regulations should at least try to follow them. May all think of my words over and over again!"

Among the numerous other inscriptions upon walls, lintels, and tablets, we would only mention one. It is composed of four characters, but when verbally translated it teaches volumes of truth in practical life: *Yü T'ien Ti T'san*, "with the heavenly earthly is mixed." However, the Confucianist sees a very different meaning in this inscription. He reads from it: "The virtue of Confucius is equal to heaven and earth."

In a little room in front of the Confucian Temple is enshrined a tutelar god. Upon inquiring why this wayside shrine was in this unusual place, we were informed that this room had become noted for the remarkable success which all students who had studied here had met with in their examinations—all having taken high degrees. In consequence of this inexplicable favor conferred upon all who studied here, there was a great rush every season for its occupancy, and when quarreling and even murder ensued, it was relegated to an idol, and no student was allowed to study there. The propitiousness of this room is ascribed to an idol—a literary god—standing in a little pavilion across the brook. This idol is facing the college and is holding a pencil in his right hand, and this pencil points directly to that room, and guides the pen of the favored student to supernatural success.

Another evidence that these Confucian scholars are far from being free of the belief in idols and images is clearly seen in the Confucian Temple located in the midst of the college buildings. In the main hall of this building are large images of Confucius, Mencius, and fifteen of the most famous disciples of Confucius. And what is not seen even at a Buddhist temple there are here four gods standing on the top of the roof, two looking northward, and two southward. In the room in front of the grotto is also a large image of *Chü Fu-tsz*.

"Scarce can the classic pilgrim sweep free
From fallen architrave the desert vine,
Trace the dim names of their divinity;
Gods of the ruined temples, where, O where are ye?"

* * *

THE MANDARIN'S GRAVE

Fifteen *li*, or about five English miles, southeast of Kiukiang, near the foot of the *Lü* Mountains, and on the southwestern slope of the Oak Tree Hill, is the grave of the celebrated *Chou Fu Tsz*, called in Chinese *Lien Chi Mu*, *Lienchi* being the name of his birthplace. To foreigners this place has become a beautiful spot for an afternoon excursion; but to the more devout Chinese it is a sacred sanctuary. If beauty of scenery and balmy air can add anything to the peaceful repose of departed spirits when they see their "mortal coil" surrounded by such lavish gifts of nature, then *Chou Fu Tsz* can certainly have nothing to regret for having chosen this location. Sheltered from the northern winds, nestled in a little amphitheater-like valley, surrounded by huge trees of many centuries' growth, with the five thousand feet mountain peaks looming up into the clouds, and the Lotus Flower Peak near by, what more could immortal shades desire? Beautiful as it is by nature, the æsthetic taste of man has added much to its picturesque harmony.

Chou Fu Tsz was a native of Hunan, born in *Lienchi* in the *Sung* Dynasty in the year 1017 A.D. He was commonly known as *Chou Tsz* and spoken of as *Lien Chi Sienseng*, the gentleman from Lienchi. When he was but a child his father died, and his mother was so poor that she brought him to her brothers, whose family name was *Chen*. When his mother died, she was buried by the side of her brothers' graves. Forty-four years later these graves were destroyed by a flood, and *Chou Fu Tsz* removed his mother's remains to their present resting place. Two years after this he died, and was buried on the left side of his mother's grave.

In the year 1488 a prefect of *Kiukiang* named *Tung* built an ancestral hall near by and endowed it with real estate property. Fourteen years later a Literary Chancellor named *Shao Pao* invited the members of the *Chou* family to worship at this hall.

Within the amphitheater a mound is raised. On top of this mound are the graves, apparently under one cover, shaped like a tortoise. In front of this tortoise-shaped cover five tablets are erected. The middle and largest one is dedicated to his mother, who received the posthumous title usually conferred upon native women of "*Taichün.*" At the upper end of this tablet the two words *Tao Ma* are engraved (The Source of the Doctrine).

To his mother's left is a tablet containing an inscription written by *Peng Yü-lin*, the celebrated admiral of the *Yangtze:* "This is the grave of the

Ancient Worthy *Lien Chi Chou Tsz* of the *Sung* Dynasty, who received posthumous honors and was named *Yuen Kung*."

To the left of this is another tablet containing an engraving of *Chou Fu Tsz*, without date or name of author. How near it resembles the worthy sage no one can tell.

The tablet to his mother's right is in honor of his two wives. His first wife was of the family *Leo*, the second of the family *Chih*. It is, however, worthy of mention that they were not his wives contemporaneously. It is believed that polygamy was less popular in those dark mediæval days than it is now.

On the tablet to the extreme right a beautiful vine is artistically carved into the side and upper part of the stone. Under the vine a boy and girl are standing. They are called *Chin Tung* and *Yü Nü*, and are supposed to be the beings who serve as guides to departed spirits over the No-Alternative Bridge—in Purgatory—which all spirits must cross.

Back of the grave is a wall built in the shape of a horseshoe, reaching round to the front. In this wall three stone tablets are set. The middle and larger one was erected by *Peng Yü-lin*, and contains the following inscription: "Southeast of *Kinkiang*, below the Lotus Flower Peak, is the grave of *Lien Chi Sienseng*. In the spring of the fifth year of *Hsien Fêng* (1856 A.D.), I, with *Li Se-ping*, brought soldiers to *Kiukiang*. When we had time we came to this grave to worship. This place was*Chou's* own choice, and his mother's grave is in the middle. When he died he was buried to his mother's left, and his wives to her right. We found their graves badly dilapidated, so we bought stones and had them repaired. A government student, a corporal, and a descendant of the twenty-second generation of *Chou's* family managed the repairs. In one month the work was completed.

"That we have had such men as *Chou* is proof that the doctrine of the ancient Sages still flourishes. If after Confucius and Mencius no other holy men had come, the people under heaven would have followed after temporal honors and riches, and have become idlers like Buddhist priests, none searching the Doctrine of the Ancient Worthies. During the two turbulent periods (from 420 to 554 A.D.) when six dynasties rose and fell (and from 907 to 960) when five dynasties followed in rapid succession, the calamities of the people were great. All because the people of those times did not know or follow the Doctrine of the Ancients. But at last a holy man was born in *Hunan*—*Chou Fu Tsz* came. Having no one to instruct him, he devoted himself to private study and meditation, and then wrote the *Ting Shu*, a calendar advising the people how to do and act according to the Doctrine of the Ancients. Then the people began to understand the teaching of Confucius and Mencius. His instruction was profoundly deep, and also practical for the moral improvement of the people. As he understood the doctrine, others also were kept from error. The teaching in the Kingdom was corrected, the Government again began to flourish, and the evil in the land ceased to prevail. If all would practice the doctrine of benevolence and righteousness as *Chou* did, applying it to their person, virtue would be secured, and in the affairs of life peace assured, and there would be culture among the officials and prosperity among the people. To save from calamity is as if saving from disease. If the constitution is sound and healthy the evil influences of disease can not come upon man. Thus *Chou's* doctrine was more and more appreciated, as he influenced his times by thinking of and teaching refinement."

On the tablet to the left of this is the following inscription in bold characters: "The Grave of the *Nankang* Prefect *Lien Chi Sienseng*. Erected first year of *Chia Ching* (1522 A.D.) by *Kao Yao-ching*, Prefect of *Nank'angfu*."

But the tablet to the right attracts the attention of the visitor above everything else to be seen here. It is called the "*T'ai Kih Tao*" (the Chart of the Absolute). This inscription is intended to set forth in picture and word the cosmogony of *Chou Fu Tsz*. A satisfactory translation of this inscription is not an easy task. As very few of the native scholars know anything of the philosophy it purports to advance, a foreigner is sometimes obliged to resort to conjecture, hence we can not claim this to be more than a free translation: "From the *U-kih* the Infinite to the *T'ai kih* the Absolute. The Absolute moved and brought forth light, the male principle. Again it moved with greater rapidity and then stood quiet, which brought forth darkness, the female principle. Then the Absolute again moved and the movement and repose were denominated '*Yang*' (light) and 'Ying' (darkness). In the Eight Diagrams the symbol of light, or positive principle, consists of a single line, thus —. The symbol of darkness; negative principle consists of a broken line, thus – –. Then light united with darkness and brought forth the five elements: fire, water, earth, metal, and wood. The five vapors—rain, sunshine, heat, cold, and wind—appeared and kept in harmony with the seasons. These elements contained darkness and light caused by the Absolute; and the Absolute also contained the five vapors. The Infinite united in deepest mystery with the five elements and the five vapors, and they firmly congealed.

"The Way of Heaven caused the male principle. The Way of Earth caused the female principle. These two acted upon each other and produced the innumerable living creatures. But man endowed by the grace of heaven, the most sagacious, has received a body that contains the functions of five senses and wisdom that enables him to distinguish good from evil. The holy man controls all these in himself and can not be easily disturbed, thus establishing a perfect man. Hence the virtue of the holy man is as great as heaven and earth, and his wisdom as clear as the sun and moon; and as the seasons are regular in their succession, so there is order in all that he does. He knows as if by intuition how to conform his actions so as to obtain the good and avoid the evil. He practices the doctrines in all benevolence and righteousness. As it has been said, if one gives thought to his beginning and end, he can know of life and death. Thus the Doctrine of the Changes is perfect above all other doctrines."

The Doctrine of the Changes referred to in this inscription is the "*I King*" (Canon of Changes), containing a fanciful system of divination from the combination of diagrams, mysterious in the extreme, called the Eight Diagrams. That this book was extant as far back as history can lead us, there can be no doubt. In the Confucian Analects VII, 14, Confucius says: "If some years were added to my life I would give fifty to the study of the '*I*' (Canon of Changes), and might then escape falling into a great error." Such was the faith of the great Sage in the Doctrine of Changes. Hearing of a man here in Kiukiang who is reported to understand the mysteries of the "*I*," I sent word to him for a little explanation of the Chart of the Absolute, and he replied, "If the gentleman who desires this information is willing to spend a lifetime in studying the '*I*,' then he will understand it." As the prime of my days is passed, I will not try to master this mysterious monument of antiquity.

Approaching the grave from the entrance, there are two tablets standing below the mound, one to the right and the other to the left. The one to the right is a small moss-covered slab containing the following inscription:

"*Chou Fu Tsz* deduced his philosophy from the Eight Diagrams, which the Emperor *Fu Hsi* discovered on the back of a tortoise. The philosopher *I Loh* is said to have perfected this doctrine. The hills are bare, and man is dead. So the stream vainly flows. For whom is the grass over the grave so beautiful? In the evening it spreads over the ground, and the chilly air makes us sad. (Signed) *Li Pa-yang*, Prefect of Kiukiang, 1573 A.D." The Emperor *Fu Hsi* mentioned in this inscription is supposed to have lived 2838 B.C., being the first of the Five Legendary Emperors. He is claimed to have been of miraculous birth, a native of *Shan si*.

To the left stands a larger tablet, erected by *Peng Yu-lin*, Admiral of the *Yangtze*, with this inscription:

"*Lienchi Chou Tsz* is our *Hunan*, *Taochou* man. His grave is below the Oak Tree Hill. In the seventh year of *Kwang Hsü* (1882), when I was Admiral of the *Yangtze*, I came to *Kiukiang*with the *Chentai* and magistrate of *Hu-kou*, and we worshiped at the grave. When I saw that our previous repairs were not substantial, I gave money and ordered the *Chentai* to repair it and build the horseshoe-shaped wall back of the grave, and make a broader entrance. I examined the historic annals of the *Sung* Dynasty, and found that *Chou Fu Tsz's* mother's grave, originally at *Renchou* in *Hunan*, had been destroyed by a flood, and *Chou Fu Tsz* had begged the Prefect of *Nank'ang* to have his mother's grave removed to this place. After this he himself became Prefect of *Nankang*, and built a home at *Lien hwa fung*. Two years later he died and was buried to his mother's left. His wives are to her right. This can be seen in the official annals of *K'anghsi*.

"During the *Sung* Dynasty this place was called *Tehhwa* Village, with an altar called the Clear Water Shrine. In the year 1215, eighth year of the Emperor *Chia Ting* of the Southern *Sung* Dynasty, a minister of state named *Wei Liao-wang*, passed here and worshiped. When he arrived at the palace in Peking he begged of the Emperor to grant *Chou* posthumous honors. This was done, and *Chou* received the name *Yuen Kung*. In the *Ming* Dynasty, during the reign of *Hung Chih* (1490 A.D.), the Prefect of *Kiukiang* built an ancestral hall near the grave, and named it *Yuen Kung Chou Siensing*. So I now write with due honors: 'The Ancient Worthy *Yuen Kung Lienchi Chou Fu Tsz Mu*.' This is in accord with ancient propriety. But to repair his grave and worship him is of no benefit to any one, if he does not receive his doctrine. If all who pass here will honor him and follow his example, then his teaching will not have been in vain.

"Now, since the work of repairs is completed, I write these words of commendation, that the chief points of his life may be held in remembrance. All who have aided in these repairs shall have their names inscribed on the opposite of this tablet.

"Done in the tenth year, third moon of *Kwang Hsü* (April, 1885). *Peng Yu-lin*."

About two miles northwest of the grave there was a school. *Lienchi Chou Fu Tsz* loving the sight of the mountains and the fresh water, had a home here. In the year 1850 there was a great flood in the *Yangtze* Valley, and this home perished, and with it the last material remembrance of this noted man disappeared; but his name will be held in high esteem by his countrymen for ages to come.

The classic pilgrim who desires to visit this shrine will find it on his way to *Kuling*, five *li* from *Shihlipu*, turning to the left at this village.

LUNG-HU SHAN—THE DRAGON-TIGER MOUNTAIN

Wherever in China this mountain is mentioned, whether in *Kiangsi* or in the most distant province, every one knows what it stands for. Just as much as Rome is known to all Catholics to be the home of the Pope of the Roman Catholic Church, so all Chinese know that the Pope of the Taoist Religion has his residence here. It was to this mountain retreat that *Chang Taoling*, the first Pope of the Taoist Religion, was directed in the first century of the Christian era. He had retired into seclusion in the mountains of Western China, devoting himself wholly to meditation and the study of alchemy. From the hands of *Laotze*, the founder of the Taoist Religion, who had lived six hundred years before him, he supernaturally received a mystic treatise by which he was enabled to compound the elixir of life. When manipulating this elixir of the dragon and tiger he met a spirit who said to him, "In the *Pesung* Mountain is a stone house where the writings of the Three Emperors and a liturgical book may be found. By getting these you can ascend to heaven, if you pass through the discipline which they enjoin." He came. He dug and found them. By means of these he was instructed how to discipline himself for a thousand days, and was then able to leave his body and walk among the stars and fight with the king of demons, divide mountains and seas, and command storm and thunder to obey him. All the demons fled before him, leaving not a trace of their retreating footsteps. On account of the prodigious slaughter of demons by this hero, various divinities came with eager haste to acknowledge their faults.

The *Pesung* Mountain is a harp-shaped section of the range that runs from northeast to southwest, separating the *Kiangsi* province from *Fukien.* Between this mountain and *Shan Sing Kung*, the Pope's resident town, a river as clear as crystal is clamoring for the *Poyang* Lake, where its pure water is soon tainted with the impurities of that body.

The latter years of the mystic earthly experience of *Chang Taoling* were spent on the Dragon-Tiger Mountains, and it was there, where at the age of 123 years, after having compounded and swallowed the grand elixir, that he had gained power to ascend to heaven to enjoy, the bliss of immortality.

Six hundred years before this (B.C. 604) *Laotsze*, the founder of the Taoist Religion, was born in the province of Honan. *Lao Tsze* means Old Boy, and doubtless was given him because at his birth his face is said to have been wrinkled and his hair gray like an aged man of seventy. Very little is known of his early life. He preferred solitude and quiet, and early withdrew from the busy haunts of man. When Confucius was quite a young man and in search of the *Tao*—the word—he came to the aged philosopher and said, "I have sought for the *Tao* for twenty years."*Laotsze* replied, "If the *Tao* could be offered to men, there is no one who would not be willing to offer it to his prince; if it could be presented to men, everybody would like to present it to his parents; if it could be announced to men, each man would gladly announce it to his brothers; if it could be handed down to men, who would not wish to transmit it to his children? Why then can you not obtain it h This is the reason. You are incapable of giving it a resting place in your heart." The young philosopher may not have relished or fully understood the instruction he received, for he came back to his disciples and said, "To-day I have seen *Laotsze*, and I can only compare him with the Dragon." Even to us he seems mysterious as the mystical animal; but we can not read the sayings of this pagan philosopher without feeling that there is a vital, restful strength and wonderful spirit manifested here that does not exist in other Chinese literature. Mystical and apt spiritual simplicity and universal wisdom is here expressed. It seems to come from a heart that has been touched by divine revelation. A few quotations from his wonderful work, the *Tao Teh King*, convinces one that he must have been not only a great and good man, but an inspired man. He doubtless had some conception of Him who was promised by the Prophets. Note these quotations, and compare them with what the beloved disciple of Christ said of the Logos: "Its name may be named, but it is not an ordinary name. Its nameless period preceded the birth of the universe. Having a name it is the mother of all things. The *Tao* is full, yet it operates as though not self-elated. In its origin it is, as it were, the ancestor of all things. I know not whose offspring it is. Its form existed before God was (by the term "God," as used in Chinese literature, is understood the First Ruler, or the Highest Ruler). It is mysterious, recondite, and penetrating. Pellucid as a spreading ocean, it yet has the semblance of permanence. There was something formed from chaos which came into being before heaven and earth. Silent and boundless it stands alone, and never changes. It pervades everything, and may be called the mother of the universe. I know not its name, but its designation is *Tao*. Heaven is *Tao*, and *Tao* survives the death of him who is the embodiment of it, living on unharmed forever. The*Tao* of heaven never strives, yet excells in victory. The *Tao* of heaven resembles a drawn bow. It brings down the high and exalts the lowly; it takes from those who have superfluity, and gives to those who have not enough. The great *Tao* is all-pervasive; it may be seen on the right and on the left. All things depend upon it and are produced by it; it denies itself to none. With tenderness it nourishes all things, yet claims no lordship over them."

Here is where *Chang Taoling*, at the age of one hundred and twenty-three, ascended to the heavens to enjoy the bliss of immortality, and in this temple is his image in life size, where all the popes have worshiped.

The philosopher *Huai Nantsze*, an adherent of Lao Tsz's philosophy, commenting on the *Tao*, described it thus: "*Tao* is that which covers heaven and supports earth; its height can not be measured, nor its depth fathomed; it enfolds the universe in its embrace. The *Tao* reaches upward to heaven and touches the earth beneath; it holds together the universe and the ages, and supplies the three luminaries with light. It is by the *Tao* that mountains are high and abysses deep; that beasts walk and birds fly; that the sun and moon are bright and the stars revealed in their courses; that the unicorn roams about and the phoenix hovers in the air. *Tao* is the beginning and end of the visible creation. *Tao*, in its sublimest aspect, regards itself as the author of creation, the power which completes, transforms, and gives all things their shape. All-pervading and everywhere revolving, yet can not be sought out; subtle and impalpable, it yet can not be overlooked. If it be piled up it will not be high; if it be added to it will not increase; if it be deducted from it will not diminish. Shadowy and indistinct, it has no form. Indistinct and shadowy, its resources have no limit. Hidden and obscure, it reinforces all things out of formlessness. Penetrating and permeating everywhere, it never acts in vain. Utterly non-existent, *Tao* is yet ever ready to respond to those who seek it."

Let one more quotation of *Laotsze* suffice. It is his conception of the ideal man, and I would pronounce it the brightest gem in pagan philosophy. "The ideal man recompenses injury with kindness." This is equal to the crowning glory of Christianity, return of good for evil.

That this great wizard chose this locality in those early days, and all succeeding popes dwelt here, is not surprising. The scenery is prodigiously picturesque, indeed, it is enchanting. Doubtless those children of nature in their imagination saw the mountains leaping like lambs. I have traveled in many lands, and have seen beautiful natural scenery, but nowhere have I ever seen a place where nature has been so lavish in bestowing her charms than in this mountain retreat. Nature truly has done her part well. But what contrast do we find when we inspect the works of man!

At the foot of the Dragon-Tiger Mountain, on the southeastern slope stands a temple, at the place where *Chang Taoling* is supposed to have ascended the throne on high. This temple, at one time doubtless a most costly structure, and though housing the image of such an honored personage and the place where the chief of a great world religion worships, is in a most dilapidated condition, the rendezvous of beggars and bats, in charge of a lonely old woman whose husband had died a year before, but in his coffin was yet standing in her living room. A few copper cash soon relaxed her stern attitude, and she freely permitted us to visit the sacred shrine with the one reserve that we do not photograph the altar and the image, lest we carry away his spirit. No persuasion or offer of money would induce her to grant us that liberty.

To *Chang Taoling* is attributed the honor of having invented the charms which in all parts of China are seen posted on door lintels, and are believed to drive away demons and all invisible malevolent beings. A proof of the dominating idea of popular Taoism throughout the land.

But the most renowned temple is about five miles northeast of this mountain, near to *Shan-sing-kung*, the town where the pope has his palace. Here, in a beautiful amphitheater-shaped valley, is a large group of temples. In these temple grounds are numerous stone tablets bearing inscriptions of ancient writers. These descriptions, however, are so defaced by the tooth of time that the deciphering of them is almost impossible. The center of this group is the Chief Hall, containing a colossal image of *Yü Hwang-ti*—the Pearly Emperor. He is emperor of all immortals. All the power of heaven and earth are supposed to be in his hands. Nothing is said of the time when he was on earth. His birthday and heaven's birthday are the same. To the Taoist worshiper he is the Supreme Ruler of the Universe. Dethrone him and Taoist image worship is demolished. To him and before his altar all the popes of the Taoist religion have prostrated and done homage.

In February, 1910, it was my privilege to visit this modern Wizard *Chang Tientsze*, the Heavenly teacher, who is a direct descendant of *Chang Taoling*, and who can boast of an unbroken lineage of sixty-two generations. To reach this mountain retreat the *Poyang* Lake had to be crossed as far as *Yao Chou-foo*, a distance of about ninety miles. This was done in a little steam yacht. One of our theological students served me as photographer. At *Hu Keo*, the Mouth of the Lake, an official joined us in our little compartment. We soon became acquainted with each other. When he had learned that we came from William Nast College, he became interested in our object of traveling at this unfriendly season. I told him because it was New Year's vacation, and that I had long since had a desire to see the Pope of the Taoist Religion, the man who is believed to have such magical power to exorcise demons and malevolent spirits. A forced smile flit across his stoical face when he replied, "In former years we did believe that he could do this, but very few people believe this now." "However," he continued after a little thought, "there is something remarkably strange about this man; he can control the lightning and thunder." Strange combination of doubt and faith. The only explanation seems that the educated official, as a Confucianist, is a materialist, and as such does not want to admit belief in the absurd popular idea of evil spirits wandering through the air disturbing the public tranquillity, but does not hesitate to admit his belief in the power of nature and the Pope's ability to subdue and control it. His unbelief in the one and his belief in the other did not, however, lessen my curiosity to meet the great Magician, the Chief of one of the great religions of the world.

Upon my arrival after four days of traveling, on the evening before the last day of the year, I sent my card to the Papal Residence, and asked for an interview the next day. Inquiry was made as to the purpose of my visit and my official relation. I replied that I was holding no official relation to my country, and had no official business, but was a humble missionary and had come from a far country, and it being the last day of the year I wished to see the great man and offer him my New Year's congratulations. That seemed quite sufficient reason to admit a stranger from a strange land, and arrangements were made for the interview on the next morning. In due time I made my appearance, when the center doors, dividing the departments, were thrown open wide, the doorkeeper walking ahead holding my card in his outstretched hand. Having passed through four divisions we reached the official palace. Here we were ushered into a well-furnished reception room where, in a few moments, his excellency appeared. He is a tall, handsome middle-aged man, was dressed in the ordinary costume of a high-class Chinese scholar, and most pleasant and congenial; well-informed in all things that were of vital interest to the Asiatic people; by no means a recluse. He is the husband of two wives, and father of one daughter and three sons. By all the Taoist priests throughout the land he is recognized as the Commander-in-chief of the Taoist religion, wielding an immense spiritual power in the entire Empire. His name is on every lip, and he is believed to be the vicegerent on earth of the Pearly Emperor in heaven, and as such he has power to expel demons from haunted houses. To accomplish this he wields the sword that is said to have come down to him as a priceless heirloom from his ancestors of the *Han* Dynasty. All demons fear this sword, and when the great magician wields it he can catch them and put them into jars which he seals with a charm. It is said that somewhere on the Dragon-Tiger Mountain there are many rows of such jars, holding incarcerated evil spirits. Personally I did not find these jars. The efficacy of a charm is supposed to be greatly increased by the magical gift of the pope from whom it is obtained, hence to secure his service is very expensive. His title *Chen-ren*, that is "the, ideal," "the true," "the perfect man," means a man who has the power to rule over himself and over nature. Being the chief official on earth representing the Pearly Emperor in heaven, he has the privilege to address memorials to him. His dwelling is denominated as the *Chen-ren-foo*. The ideal man's home.

Having learned that I was a missionary, he seemed the more delighted to see me. In speaking of our Church, he made a rather ambiguous remark. "I do not see," said he, "why your people and our people should not live together in pleasant harmony, for our aim is the same; we are working for the same end." At the time I could not understand his remark, and being his guest I did not pretend to question his well-meant assertion of our common aim in life. When I returned to the Home of the China Inland Mission, where a native Christian woman was doing effective evangelistic work, I told her what *Chang Tientsze* had said about our relation to each other. She immediately understood what influenced him to make this remark, and told me the following story:

A few months before a rich Cantonese woman, accompanied by a great retinue, came to the Pope for treatment. She was ailing of an incurable malady, supposed to be demon possession. The Pope soon took in the situation, and pronounced her case very difficult of treatment. So serious did he claim it to be, that nothing less than Tls. 3,000, which is equal to $1,100 United States currency, would suffice to undertake the cure. When the bargain was concluded, the treatment was begun. In the meantime a poor woman came to be treated who was suffering of insomnia. When the Pope heard of her presence he sent word that he was too busily engaged in fulfilling a contract with another person, he had no time to spare. The poor woman, however, had entered the first door and found shelter under a little roof by the side wall. The Christian Evangelist woman, having heard of the pitiable condition of the poor woman, came to see her, but being known as an evangelist worker, was not permitted to enter there and do that kind of work. The poor woman receiving no help here under the roof of the Perfect Man, went to the Temple where the Pearly Emperor had his shrine. Having received no mercy from the Vicegerent, she hoped to move the compassion of the Ruler of Heaven himself. The, evangelist woman also found her there and saw her kneeling before the great image worshiping. Approaching her with the tenderness of a mother, she said to her, "What is your trouble, and why are you here in agony?" The poor woman told the sad story of her life, her ailment, and her disappointment at the Heavenly Teacher's palace. The evangelist woman said, "Whatever you may ask of this renowned image, be assured you ask in vain; it is only wood as other wood, and can not help you. We worship the true God, the Creator of all, and His Son Christ our 'Savior, the Redeemer of all. Believe in Him and you shall be saved. I know not your ailment, but He is mighty to save and present everywhere." The woman remained in the temple and began to pray to the unknown God, of whom she had heard for the first time. That night she slept like a child, the first sleep she had had for years, sleeplessness having been her ailment. The following day the evangelist again went to the temple and found the woman calm and at rest.

She invited her to the chapel, where she heard more of the Savior who had saved her and given her rest. She was healed. The Pope heard of this, and was not a little discomfited, thinking the evangelist woman had learned his art of exorcising the evil spirits, and thus healed the woman who in vain appealed to him for help because she had brought him no earthly goods. Hence he inferred that our work and his must be the same. Now I understood his remarks.

In the conversation I told him of our educational work at William Nast College in Kiukiang, and the advantage of such a learning to the young men of China. When he heard that non-adherents to the Christian religion were admitted, he concluded that one of his boys must have a Western learning. Of course we can hardly expect that the eldest son would be given this privilege, for he will be the future Pope, and a Taoist Pope with a Christian education would be like a combination of darkness and light. However, we shall consider it as a divine guidance if we succeed in getting either of the younger boys.

Before we left the town a distant relative decided to go with us as a student to the college. Thus a little beginning is made to break into this stronghold of spiritual darkness. A little light ignited here may work wonders to a benighted people.

According to Oriental custom, he returned my call on the following day. Upon this occasion he was clad in the garment of a high official of State. His Canonical robes, made of the finest gold and silver embroidery, are only worn when engaged in the functions of his official duties as the great Magician.

* * *

CONCLUSIONS

The places described in this little volume represent the centers of religious life among the three religions of the great Imperial Kingdom. The time spent in visiting these places gives one an entirely new view of the relation of the religious systems of China to each other, and the hold they have upon the people. Strange as it may be, we find them opposing each other, and yet closely interlocked in silent partnership, of which even the Government is not excluded. Of course, in an Imperial Kingdom which numbers its inhabitants by hundreds of millions, with dialects, temperament, endeavors, and beliefs so different and varied, it is expected to find the greatest possible religious tolerance. And this is here true in-as-far as the principles of the different religious systems are in harmony with the statutes of the Government. For example, the reigning dynasty gave to the religious systems which they found in vogue, when they conquered the *Mings*, full and complete liberty in all things that pertained to religion. That this policy depended more upon the force of circumstances than their liberality and the exalted opinions they held regarding the *Mings*, can not be questioned; for the *Manchus*, being a rough, uncultured race, with no definite belief, they accepted the existing religious systems of China. Their Emperor was made Chief Priest of Confucianism, which was the State religion of the conquered provinces, and the Buddhists, Taoists, Mohammedans, and Jesuits were all recognized and tolerated. In the first instance this tolerance doubtless was due to the indifference towards all systems of faith. They found the Chinese a quiet, industrious people, and as long as they were loyal and obedient they were permitted to believe what they pleased and worship whom they chose; but whenever the Imperial Government discovered combinations in any religious system that might threaten the peace and welfare of the throne, then the Emperor meted out the cruelest vengeance upon the supposed transgressors.

This religious indifference upon the part of the Government has not existed without showing signs of weakness and vacillation. Notice, for example, the policy pursued by the most enlightened and energetic Emperor of the present dynasty, *Kang Hsi.* As Emperor he was the nation's High Priest, and worshiped *Shangti*—the High Ruler, at the Altar of Heaven. And in his sacred edicts he pronounced the Buddhists and Taoists as dangerous heretics, while at the same time he had Buddhist and Taoist temples renovated and repaired at his own cost, and worshiped in them. Such examples of duplicity upon the part of Emperors have not been without effect upon the nation. The inevitable result of trying to adhere to three opposing systems of religion is seen in the gradual decay of all religious life among the higher classes. The constant interchanging has produced spiritual stagnation in high places. To-day the great mass of the people choose at will of these systems, and when they have made their choice they little think of the admixture of their temporary elective system. Ask an ordinary Chinese to which of the three religions he adheres, and he will be unable to give an intelligent answer. If he has had some education and considers himself a scholar, then he will at once reply that he is a Confucianist; but it is quite possible that the next day you will find this gentleman burning incense in a Buddhist or Taoist temple, or you may find a priest at his home reading mass for some departed member of the family. Even among the officials, who are all supposed to be strictly Confucianists, there are many who in their official costume enter the temples and worship the idols.

With such lack of conviction, is it strange that Confucianism and its opposing systems, Buddhism and Taoism, have long since passed their zenith and are rapidly approaching their dissolution? The unpalatable principles of the "Great Wise Man of China," however beautiful and good, and however much good they may have been to the generations of past ages, they can not change the heart of their people nor satisfy their present aspirations. Confucianism has been tested for twenty-five centuries, during which time it has added nothing to the spiritual life of the Chinese people; while its opponent systems, which had some semblance of spiritual life in their beginning, have always been a thorn in the flesh to the highly cultured Mandarin, and often a ridicule to the people. But even in the face of all this decay and the obvious advancement of Western ideas, there are no signs of uneasiness noticeable upon the part of these religions about the wonderful progress of Christianity. And we believe it quite safe to say that no serious obstacle will be placed in its way as long as its propagation gives no cause of suspicion through political interference. No thought is of greater importance to-day in the propagation of the Christian religion than the maintenance of the integrity of the Imperial Kingdom, and carefully avoiding all interference with political questions. The stigma that once rested upon the missionaries has been removed. The Protestant missionaries are no longer looked upon as emissaries of their governments. The officials have come to see that the dominant desire of the missionary is to see the cause of Christ advanced and order and loyalty promoted.

It is indeed pathetic to see such a great mass of human beings so wholly absorbed in things temporal and yet unable to rise from the crushing burden of poverty and misery. If hard labor and diligence could have brought relief they long since would have risen to wealth and happiness. But vastly more pathetic is it to see them in their efforts to satisfy the soul's longing, and in their penury spend millions of dollars in making pilgrimages to far distant shrines, bringing sacrifices to the dumb idols, involving self-sacrifice of which in Christian lands we have no conception. It must be terrible to be so poor that life no longer is life, but only an existence; but infinitely more terrible to be fettered down by superstition to gods that need to be appeased by such sacrifices, and yet bring no blessing. It is hard to be a heathen! Surely, if God could be found by mere searching for blessings and bringing sacrifice, these people would long since have found Him and rejoice in Him. But their systems to which they trusted, and now hoary with age, have not lifted them upward nor led them to the truth, but estranged them farther and farther away from the truth, and led deeper and deeper into superstition and misery. The only joy these visits brought to me was the convictions of thereby becoming better prepared to guide some of these wandering millions to find the way to Him who is the Light of the world and rejoice in the living God.

---

* * *

Made in the USA
Monee, IL
09 February 2024

53144432R00038